Campylobacter

Edited by Guillermo Tellez-Isaias and Saeed El-Ashram

Published in London, United Kingdom

IntechOpen

Supporting open minds since 2005

Campylobacter
http://dx.doi.org/10.5772/intechopen.92910
Edited by Guillermo Tellez-Isaias and Saeed El-Ashram

Contributors
Xiaolun Sun, Ying Fu, Hong Wang, Tahrir Alenezi, Ayidh Almansour, Zhenquan Jia, Hongsheng Huang, Manuel Mariano Garcia, Taghrid Istivan, Mohsina Huq, Ali Raza Ishaq, Ayesha Noor, Laila Jafri, Faiza Jabeen, Rehana Rani, Bushra Hafiz Kiani, Nosheen Akhtar, Zeeshan Javed, Tahira Younis, Fatima Jalal, Guillermo Téllez, Saeed El-Ashram, Cheng He, Victor Manuel Petrone-Garcia, Musafiri Karama, Beniamino Cenci-Goga, Luca Grispoldi, Reem Alajmi, Abdulaziz S. Alouffi, Norma L. Calderón Apodaca, Adriana del Carmen Gutiérrez-Castillo, Leopoldo Henri Paasch-Martínez

First published in London, United Kingdom, 2022 by IntechOpen
IntechOpen is the global imprint of INTECHOPEN LIMITED, registered in England and Wales, registration number: 11086078, 5 Princes Gate Court, London, SW7 2QJ, United Kingdom
Printed in Croatia

British Library Cataloguing-in-Publication Data
A catalogue record for this book is available from the British Library

Additional hard and PDF copies can be obtained from orders@intechopen.com

Campylobacter
Edited by Guillermo Tellez-Isaias and Saeed El-Ashram
p. cm.
Print ISBN 978-1-83968-975-8
Online ISBN 978-1-83968-976-5
eBook (PDF) ISBN 978-1-83968-977-2

For EU product safety concerns:
IN TECH d.o.o., Prolaz Marije Krucifikse Kozulić 3, 51000 Rijeka, Croatia, info@intechopen.com or visit our website at intechopen.com.

Meet the editors

Guillermo Tellez-Isaias received his DVM and MS in Veterinary Sciences from the National Autonomous University of Mexico (UNAM), and his Ph.D. from Texas A&M University. He worked as a professor at UNAM for sixteen years, eight as head of the Avian Medicine Department, College of Veterinary Medicine. Dr. Tellez was president of the National Poultry Science Association of Mexico and is a member of the Mexican Veterinary Academy and the Mexican National Research System. Currently, he works as a research professor at the Center of Excellence in Poultry Science, University of Arkansas. His research is focused on poultry gastrointestinal models to evaluate the beneficial effects of functional foods to enhance intestinal health and disease resistance.

Dr. Saeed El-Ashram is a professor at Foshan University, China, and Kafrelsheikh University, Egypt, and a research professor at Zhaoqing Dahuanong Biology Medicine Co., Ltd., China. Dr. El-Ashram's research focuses on parasitic diseases. He has more than 100 journal publications to his credit. He is currently an academic editor and reviewer and holds several registered patents. The primary focus of his research is to understand how the animal immune system recognizes and responds to parasitic infections with and/or without a microbial community. Some are the causative agents of significant diseases in humans, such as toxoplasmosis, cryptosporidiosis, alveolar echinococcosis, and fascioliasis. Others are a substantial financial burden to food producers because of the effects these parasites have on domestic animals, for example, coccidiosis and cryptosporidiosis (livestock and poultry).

Contents

Preface

Campylobacter is well recognized as the leading cause of bacterial food-borne diarrheal infections worldwide. Symptoms can range from mild to severe conditions, including permanent neurological symptoms. Recent studies have shown that 31% of Guillain-Barré syndrome cases, a neurologic disease that causes ascending paralysis, are attributable to *Campylobacter* infection. *Campylobacter* is a thermophilic bacterium. Generally, the strains of *Campylobacter* are apathogenic in poultry, although newly hatched chicks and turkeys may develop transient diarrhea following *C. jejuni* infection. Modern intensive poultry production favors the introduction of disease into commercial growing units, resulting in intestinal colonization during the second to fourth weeks. The organism is carried in the intestines of many wild and domestic animals; hence, routes of infection in commercial poultry include contaminated fomites, infected water supplies, rodents, insects, and free-living birds. Intestinal colonization results in healthy animals as carriers, and epidemiological data suggest that contaminated products of animal origin, especially poultry, contribute significantly to Campylobacteriosis. Consequently, the reduction of raw poultry contamination has a significant impact on reducing the incidence of infection. Contamination of poultry products occurs both on the farm and in processing plants. Routine procedures on the farm, such as feed withdrawal, poultry handling, and transportation practices, have a documented effect on *Campylobacter* levels at the processing plant. At the plant, defeathering, evisceration, and carcass chillers have been reported to cross-contaminate poultry carcasses. The high frequency of *Campylobacter* spp. transmission from poultry to humans has prompted scientists to consider and create alternative intervention strategies to control the pathogen in poultry production since excessively high numbers of *Campylobacter* (often $> 10^8$ cfu/g of poultry intestinal material) potentiate high numbers of the organism in the processed broiler carcass, with an increasing consequent human health risk. Interventions during poultry production portend the most excellent opportunity for reducing the risk of disease. However, amelioration of infection by applying improved hygiene standards and decontamination approaches, such as washing carcasses and applying chemical disinfectants and gamma irradiation, can reduce the prevalence of *Campylobacter* contamination in poultry meat. This book assesses the significance of *Campylobacter* as a food-borne pathogen and consolidates recent advances in isolation, identification, role of immune responses and microbiota, new perspectives, and novel control strategies.

The editors express their sincere appreciation to all the authors who contributed to this book for their hard work and dedication, as well as to the IntechOpen editorial team for allowing us to complete this project.

Guillermo Tellez-Isaias
Department of Poultry Science,
University of Arkansas Agricultural Experiment Station,
Fayetteville, AR, USA

Saeed El-Ashram
College of Life Science and Engineering,
Foshan University,
Foshan, Guangdong Province, China

Faculty of Science,
Kafrelsheikh University,
Kafr-El-Sheikh City, Egypt

Chapter 1

Introductory Chapter: The Significance of *Campylobacter* as Foodborne Pathogen

Saeed El-Ashram, Cheng He, Guillermo Tellez-Isaias, Victor M. Petrone-Garcia, Musafiri Karama, Beniamino Cenci-Goga, Luca Grispoldi, Reem Alajmi and Abdulaziz S. Alouffi

1. Introduction

1.1 *Campylobacter*

Campylobacter is related to *Arcobacter*, *Helicobacter*, and *Wolinella* and is classified as proteobacteria in the order Campylobacteriales and the family Campylobacteraceae [1, 2]. Sebald and Veron (1963) were the first to propose the genus *Campylobacter*, which now has 27 species and 8 subspecies [3]. *Campylobacter* is gram-negative bacteria that are curved, spiral, or rod-shaped and do not produce spores. They are tiny (0.2–0.9 μm in width and 0.2–5 μm in length). With the exception of the nonmotile *Campylobacter gracilis* and *Campylobacter showae*, which possess numerous flagella, most species have a corkscrew-like movement through a single polar flagellum or bipolar flagella [4]. This genus of bacteria is chemoorganotroph, meaning they get their energy from amino acids and tricarboxylic acid cycle intermediates. They have high dietary requirements and need a variety of nutritional settings, including anaerobic or micro-aerobic conditions. Some species of this genus are considered commensal organisms. However, they are linked to a broad range of illnesses in animals and humans. The three species of the genus, including *Campylobacter jejuni*, *Campylobacter coli*, and *Campylobacter lari*, which are usually referred to as thermophilic species, account for the bulk of human infections. *C. coli* is the second most prevalent *Campylobacter* species related to human sickness, and it is often found in pigs. *C. jejuni* subspecies *jejuni* (also known as *C. jejuni*) and *C. jejuni* subspecies *doylei* are the two subspecies of *C. jejuni*. *C. jejuni* is a pathogen that is found as a commensal in chickens and is considered a major foodborne pathogen. *C. jejuni* subspecies *doylei* is distinct from *C. jejuni* in that it does not have any animal hosts [5, 6]. Since its discovery, *Campylobacter* has been recognized as a significant human gastrointestinal pathogen globally. According to the Emerging Pathogens Institute's Foodborne Illness Risk Ranking Model (FIRRM), *Campylobacter* is the most prevalent foodborne pathogen in the United States, posing the greatest public health burden [7]. Since 2013, *Campylobacter* infection has been the most frequently detected infection in FoodNet locations, and the rate of infection seems to be rising. The poultry industry is a significant source of *Campylobacter*. *Campylobacter* is a zoonotic-causing commensal bacteria found in the gastrointestinal tracts of many wild animals

 IntechOpen

(birds like ducks and gulls), agricultural animals (cattle and pigs), and companion animals (dogs and cats) [8]. *C. jejuni* causes the majority of cases within a vast and varied collection of species, while *C. coli* causes 1–25% of *Campylobacter*-related disorders [9]. Due to occasional exposure to this disease, vulnerable populations of *Campylobacter* have been found to be concentrated in the United States and European nations [10]. Campylobacteriosis has been shown to be seasonal, with a surge in the summer months, perhaps due to an increase in flies and other vectors [11]. In flocks, the infection passes horizontally from the environment to 2- or 3-week-old chicks due to protective maternal antibodies retained in serum for a week after hatching. Then progressively reduce it until the third week is complete. *Campylobacter* infection quickly spreads horizontally among the population, with an incidence rate of 2.37 cases per day. One infected bird may spread *Campylobacter* to 20,000 chickens within a week. Numerous investigations have shown that vertical Campylobacter infection is possible. *Campylobacter* was recovered from the internal and external surfaces of eggshells, the maternal reproductive system, and rooster sperm [12]. The majority of human illnesses are caused by eating raw, infected animal products, including meat and milk, particularly chicken meat. People may be infected by drinking contaminated water, coming into contact with animals, and other environmental causes [11]. Campylobacteriosis is characterized by acute diarrhea that is often accompanied by abdominal cramping, headache, and fever [13]. Campylobacteriosis is a self-limiting illness with a latent phase of 2–5 days and a clinical duration of up to 2 weeks [14]. However, the symptoms may persist for many weeks, and in 10% of reported cases, medical intervention is necessary [15]. Guillain-Barre syndrome (GBS), a life-threatening autoimmune illness that causes peripheral neurological injury and has a death rate of 2–7% and a poor prognosis, is attributed to *C. jejuni* infection [16, 17].

1.2 Human campylobacteriosis

Recent advances in epidemiological monitoring, including the use of genetic techniques, have resulted in the detection of at least ten additional *Campylobacter* spp., other than *C. jejuni* and *C. coli*, in gastroenteritis patients; these have been designated as emerging *Campylobacter* pathogens [18]. *Campylobacter concisus* and *Campylobacter upsaliensis* have the greatest incidence among them. Furthermore, in samples obtained from southern Ireland, *Campylobacter ureolyticus* seems to have surpassed *C. coli* as the second most prevalent causal agent of campylobacteriosis. Except for ten *Campylobacter* species, all have been linked to human sickness, and infections are most often induced by the eating of infected meat, especially chicken, milk, or water, or by contact with the environment [19]. Infections with *Campylobacter* usually induce acute gastroenteritis, but they may potentially cause severe extra-intestinal sickness or long-term neurological or gastrointestinal problems (**Figure 1**, [20]).

For the most part, *Campylobacter* infections are self-limiting gastroenteritis with no long-term repercussions. *Campylobacter jejuni* is the most commonly encountered cause of *Campylobacter* gastroenteritis worldwide, followed by *Campylobacter coli*. "Emerging *Campylobacter* species, " which include *Campylobacter concisus, Campylobacter lari, Campylobacter ureolyticus,* and *Campylobacter upsaliensis*, have also been found often in patients with gastroenteritis [20]. Complicated cases of campylobacteriosis, including extra-intestinal infections and long-term sequelae, contribute significantly to the total illness burden of *Campylobacter*, even though they are very uncommon. Individuals who are immunocompromised, pregnant, or old are more likely to get extra-intestinal infections. *Campylobacter*-associated bacteremia, septicemia, meningitis, spontaneous abortion, neonatal sepsis, abscesses, soft tissue infections, cardiovascular

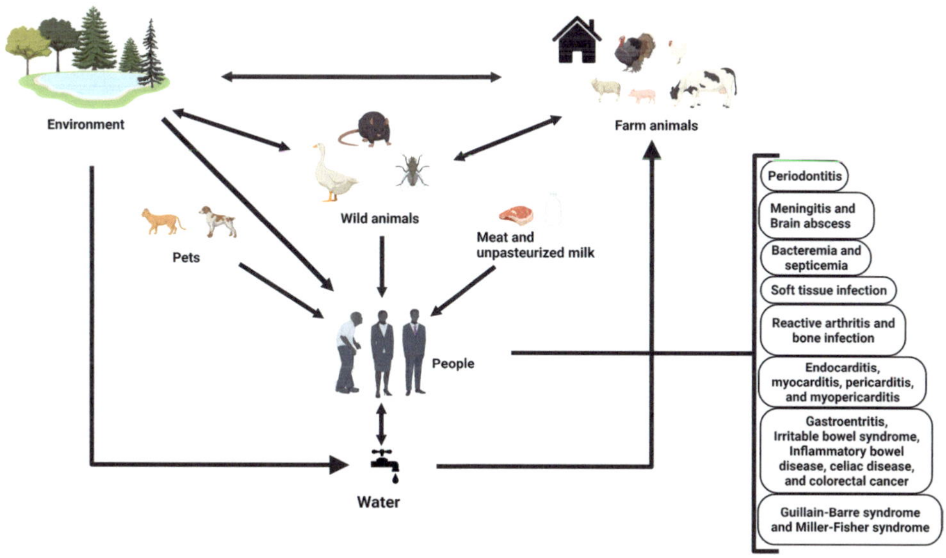

Figure 1.
Campylobacter *infections in humans: Sources and clinical symptoms. The main sources and pathways of* Campylobacter *transmission are shown on the left side of the figure. Arrows indicate transmission routes that are still under investigation. The variety of clinical symptoms of* Campylobacter *infections documented in humans is shown on the right-hand side of the figure.*

problems, and periodontal disease have all been reported. A tiny percentage of individuals with Campylobacter gastroenteritis suffer long-term post-infectious sequelae, such as Guillain-Barré syndrome (0.07%), reactive arthritis (2.86%), and IBS (irritable bowel syndrome) (4.01%) [21]. Guillain-Barré syndrome (GBS) is the most severe post-infectious consequence, with immediate flaccid paralysis and ophthalmoplegia, while Miller Fisher syndrome (MFS), a clinical variation of GBS, causes ophthalmoplegia and cerebellar-like ataxia [22]. While immunological diseases are well-known long-term effects of campylobacteriosis, there is also mounting evidence of a link between *Campylobacter* infection and chronic gastrointestinal illnesses, including inflammatory bowel disease (IBD), IBS, celiac disease, esophageal disease, and colon cancer [21, 23].

1.3 *Campylobacter* in livestock

The species *Campylobacter fetus*, which was most likely the first Campylobacter identified, is primarily responsible for the clinical relevance of *Campylobacter* infection in animals. *C. fetus* subsp. *fetal* and *C. fetus* subsp. *venerալis* are the two subspecies of the species. *C. fetus* subsp. the fetus has been found in a variety of animals, including poultry, reptiles, and humans, but is most often related to sheep and bovine miscarriage [24, 25]. Immunocompromised humans, as well as instances of neonatal sepsis and septic abortion, have all been discovered to have the pathogen. Bovine genital campylobacteriosis and infected venereal illness are caused by *C. fetus* subsp. *veneralis*, which may cause infertility, abortion, and embryo mortality [26]. *Campylobacter* is present in a broad spectrum of birds and animals as an asymptomatic colonizer and is regarded as a public health issue when identified in cattle and pets [8, 20]. *Campylobacter* spp., such as *C. jejuni* and *C. coli*, but also *C. upsaliensis*, *C. concisus*, *C. lari*, and *Campylobacter lanienae*, are often found as colonizers in the digestive tracts of chickens, pigs, and cattle. Once Campylobacter has been introduced to a flock, it may quickly spread. It usually leads to the life-long colonization of poultry [8].

1.4 *Campylobacter* in poultry

Campylobacter infection has been the most commonly found infection in FoodNet sites since 2013, and the prevalence of infection seems to be increasing. According to the Emerging Pathogens Institute's FIRRM, *Campylobacter* is the most common foodborne pathogen in the United States, causing the highest public health risk [7]. According to epidemiologic research based on molecular epidemiology, chicken may be the leading cause of human campylobacteriosis [27]. In the meanwhile, owing to its greater feeding efficiency and quicker development rate than pork and beef, poultry is one of the world's most significant animal protein resources [28]. The consumption of meat has shifted to poultry. This reflects the cheaper price of chicken compared to other meats in low-income developing nations, while it demonstrates a growing preference for white meats in high-income countries, which are easier to cook and seen as a healthier dietary option. Poultry meat is expected to account for 41% of total protein derived from meat sources worldwide in 2030, an increase of 2% from the baseline period (OECD/FAO, 2021–2030). *Campylobacter* is commensal bacteria that cause chronic infections in birds, causing little or no visible symptoms despite extensive colonization. Poultry is extensively colonized by *Campylobacter*, particularly *C. jejuni* and *C. coli*, and serves as a natural reservoir for these bacteria. *C. jejuni* infection is the most common cause of food-borne gastroenteritis in individuals worldwide [29]. When the first bird in a flock gets colonized, illness quickly spreads throughout the flock [11]. *Campylobacter* spreads rapidly throughout the flock, most likely as a consequence of fecal-oral transmission exacerbated by shared water and feed [15]. *C. jejuni* can also survive in feather follicles and pores on chicken skin at a depth of 20–30 m upon contact with its poultry host, which gives *C. jejuni* an appropriate microenvironment with little exposure to oxygen, appropriate humidity, and temperature to endure stress conditions [30]. *C. jejuni* colonization may continue throughout the broiler's lifecycle, resulting in carcass contamination at the slaughter site. Although *Campylobacter* may be detected in the mucous layer of broiler chickens' intestines, it is more often found in the cecal and cloacal crypts, where it does not attach to epithelial cells [31]. Broiler chickens may harbor *C. jejuni* at concentrations of up to 10^6 to 10^{10} CFU per gram of feces [15]. *C. jejuni* discovery in tissues other than the intestine, such as the spleen, lung, heart, and liver, shows that this virus may translocate intestinal epithelial cells and become systemic [32]. Epidemiological tests suggest that a two-log unit decrease in *C. jejuni* contamination in chicken carcasses may result in a 30-fold reduction in the risk of human infection [33]. Additionally, this shows that eradicating *C. jejuni* on farms would have a major influence on the lowering of campylobacteriosis in humans.

Author details

Saeed El-Ashram[1,2*], Cheng He[1*], Guillermo Tellez-Isaias[3*], Victor M. Petrone-Garcia[4], Musafiri Karama[5], Beniamino Cenci-Goga[6], Luca Grispoldi[6], Reem Alajmi[7] and Abdulaziz S. Alouffi[8]

1 College of Life Science and Engineering, Foshan University, Foshan, China

2 Faculty of Science, Kafrelsheikh University, Kafr el-Sheikh, Egypt

3 Department of Poultry Science, University of Arkansas, Fayetteville, AR, USA

4 Departamento de Ciencias Pecuarias, Universidad Nacional Autónoma de México (UNAM), Cuautitlan Izcalli, México

5 Department of Paraclinical Sciences, Faculty of Veterinary Science, University of Pretoria, Onderstepoort, South Africa

6 Dipartimento di Medicina Veterinaria, Università degli Studi di Perugia, Perugia, Italy

7 Department of Zoology, Faculty of Science, King Saud University, Riyadh, Saudi Arabia

8 King Abdulaziz City for Science and Technology, Riyadh, Saudi Arabia

*Address all correspondence to: saeed_elashram@yahoo.com, chenghe@cau.edu.cn and gtellez@uark.edu

References

[1] Busse H-J. 11 - polyamines. In: Rainey F, Oren A, editors. Methods in Microbiology. Academic Press; 2011. pp. 239-259

[2] Moreno-Torres IC. Campylobacter and Helicobacter. In: Reference Module in Biomedical Sciences. Elsevier; 2021

[3] Véron M, Chatelain R. Taxonomic study of the genus Campylobacter Sebald and Véron and designation of the Neotype strain for the type species, Campylobacter fetus (Smith and Taylor) Sebald and Véron. International Journal of Systematic and Evolutionary Microbiology. 1973;**23**(2):122-134

[4] Silva MF et al. *Campylobacter portucalensis* sp. nov., a new species of Campylobacter isolated from the preputial mucosa of bulls. PLoS One. 2020;**15**(1):e0227500

[5] Steele TW, Owen RJ. *Campylobacter jejuni* subsp. doylei subsp. nov., a subspecies of nitrate-negative campylobacters isolated from human clinical specimens. International Journal of Systematic and Evolutionary Microbiology. 1988;**38**(3):316-318

[6] Keener KM et al. Comprehensive review of Campylobacter and poultry processing. Comprehensive Reviews in Food Science and Food Safety. 2004;**3**(2):105-116

[7] Batz M, Hoffmann S, Morris JG Jr. Disease-outcome trees, EQ-5D scores, and estimated annual losses of quality-adjusted life years (QALYs) for 14 foodborne pathogens in the United States. Foodborne Pathogens and Disease. 2014;**11**(5):395-402

[8] Facciolà A et al. Campylobacter: From microbiology to prevention. Journal of Preventive Medicine and Hygiene. 2017;**58**(2):E79-E92

[9] Sahin O et al. Campylobacter in poultry: Ecology and potential interventions. Avian Diseases. 2015;**59**(2):185-200

[10] Ck O et al. Epidemiology of *Campylobacter jejuni* infections in industrialized nations. In: Campylobacter 3rd Edition, II. Clinical and Epidemiologic Aspects of Campylobacter Infections. 2008

[11] Gölz G et al. Relevance of Campylobacter to public health--the need for a one health approach. International Journal of Medical Microbiology. 2014;**304**(7):817-823

[12] Abd El-Hack ME et al. Approaches to prevent and control Campylobacter spp. colonization in broiler chickens: A review. Environmental Science and Pollution Research International. 2021;**28**(5):4989-5004

[13] Blaser MJ. Epidemiologic and clinical features of *Campylobacter jejuni* infections. The Journal of Infectious Diseases. 1997;**176**(Suppl 2):S103-S105

[14] Young KT, Davis LM, Dirita VJ. *Campylobacter jejuni*: Molecular biology and pathogenesis. Nature Reviews. Microbiology. 2007;**5**(9):665-679

[15] Lee MD, Newell DG. Campylobacter in poultry: Filling an ecological niche. Avian Diseases. 2006;**50**(1):1-9

[16] Jasti AK et al. Guillain Barré syndrome: Causes, immunopathogenic mechanisms and treatment. Expert Review of Clinical Immunology. 2016;**12**(11):1175-1189

[17] Mawla NN, Sultana S, Akhter N. Guillain-Barré syndrome and *Campylobacter jejuni* infection: A review. Delta Medical College Journal. 2014;**2**(1):28-35

[18] Vandamme P, De Ley J. Proposal for a new family, Campylobacteraceae. International Journal of Systematic and Evolutionary Microbiology. 1991;**41**(3):451-455

[19] Mughini-Gras L et al. Campylobacteriosis in returning travellers and potential secondary transmission of exotic strains. Epidemiology and Infection. 2014;**142**(6):1277-1288

[20] Kaakoush NO et al. Global epidemiology of Campylobacter infection. Clinical Microbiology Reviews. 2015;**28**(3):687-720

[21] Keithlin J et al. Systematic review and meta-analysis of the proportion of Campylobacter cases that develop chronic sequelae. BMC Public Health. 2014;**14**:1203

[22] Wakerley BR, Yuki N. Guillain-Barré syndrome. Expert Review of Neurotherapeutics. 2015;**15**(8): 847-849

[23] Riddle MS et al. The chronic gastrointestinal consequences associated with campylobacter. Current Gastroenterology Reports. 2012;**14**(5):395-405

[24] Wagenaar JA et al. Campylobacter fetus infections in humans: Exposure and disease. Clinical Infectious Diseases. 2014;**58**(11):1579-1586

[25] Gorkiewicz G et al. A genomic island defines subspecies-specific virulence features of the host-adapted pathogen *Campylobacter fetus* subsp. venerealis. Journal of Bacteriology. 2010;**192**(2):502-517

[26] Truyers I et al. Diagnosis and management of venereal campylobacteriosis in beef cattle. BMC Veterinary Research. 2014; **10**(1):280

[27] Agunos A et al. A systematic review characterizing on-farm sources of Campylobacter spp. for broiler chickens. PLoS One. 2014;**9**(8):e104905

[28] Smil V. Eating meat: Evolution, patterns, and consequences. Population and Development Review. 2002;**28**(4): 599-639

[29] Umar S, Maiyah AT, Mushtaq A. Campylobacter infections in poultry: Update on challenges and potential immune interventions. World's Poultry Science Journal. 2016;**72**(2):381-390

[30] Chantarapanont W, Berrang M, Frank JF. Direct microscopic observation and viability determination of *Campylobacter jejuni* on chicken skin. Journal of Food Protection. 2003;**66**(12):2222-2230

[31] Lin J. Novel approaches for Campylobacter control in poultry. Foodborne Pathogens and Disease. 2009;**6**(7):755-765

[32] Knudsen KN et al. *Campylobacter jejuni* strains of human and chicken origin are invasive in chickens after oral challenge. Avian Diseases. 2006;**50**(1): 10-14

[33] Rosenquist H et al. Quantitative risk assessment of human campylobacteriosis associated with thermophilic Campylobacter species in chickens. International Journal of Food Microbiology. 2003;**83**(1):87-103

Chapter 2

Salmonellosis and Campylobacteriosis, Emerging Zoonosis in the World and Current Situation in Mexico

Adriana del Carmen Gutiérrez-Castillo, Leopoldo Henri Paasch-Martínez and Norma Leticia Calderón-Apodaca

Abstract

Salmonellosis and campylobacteriosis are the furthermost common zoonotic infections around the world that are transferred. The spread of *Salmonella enterica* serotypes Enteritidis (SE) and Typhimurium (ST) has increased dramatically in the last 50 years due to the consumption of food contaminated and the emergence of SE and ST infections with multiple antibiotic resistance. Retrospective investigations imply an epidemiological link between people and poultry. It has been argued that farm modernization and global exports of progenitor birds have had a vital role in spreading SE and ST. On the other hand, campylobacteriosis is more common than salmonellosis in affluent countries. *Campylobacter jejuni* has been identified as the primary cause of acute diarrheal illnesses, frequently associated with animal-derived foods, particularly poultry meat. The current review examines immunological and molecular biological techniques that allow for the quick detection of asymptomatic animal carriers, as well as recent characterizations of relevant taxonomic and pathogenic characteristics of these organisms. We further urge epidemiological research to evaluate the incidence of human diseases arising from poultry eating, based on preliminary non-publisher findings implying a prevalence of salmonellosis and campylobacteriosis in Mexican poultry farms comparable to other nations.

Keywords: *Salmonella*, *Campylobacter*, poultry, epidemiology, zoonosis

1. Introduction

Animal protein is the source of a significant number of zoonosis in humans. Salmonellosis and campylobacteriosis are currently important zoonoses in industrialized countries [1].

Salmonellosis is predicted to cause about 18,000 sicknesses and 500 diseases in the USA annually [2]. In Denmark, the yearly cost of infection in humans is estimated to be USD 15.5 million. Denmark spent USD 14.1 million in a program to eradicate SE, which is considered low to USD 25.5 million estimated for losses caused by non-work and medical treatment [3]. Unfortunately, information related to the cost of foodborne illness is generally not well reported or published in developing countries [3].

In some countries, salmonellosis problems have increased 20-fold between the eighties and nineties of the last century [4].

A retrospective analysis of salmonellosis cases carried out in Norway between 1966 and 1996 suggests an epidemiological relationship between birds and humans [5]. In the United Kingdom, salmonellosis and campylobacteriosis have been found to increase between June and August, which is attributed to the lack of timely refrigeration of food being stored in refrigeration just as the ambient temperature rises and also due to the habit of consuming barbecue since the meat is not adequately cooked during the summer [2].

Over the past decade, the number of salmonellosis cases recorded in Sweden has doubled due to more infections of SE, with four *salmonella enterica* serotypes were responsible for 60% of the cases detected in that country in 2001: ST (22.1%), SE (17.7%), *S.* Newport (10%), and *S.* Heidelberg (5.9%) [6].

In 2005, three serotypes of *Salmonella* were responsible for more than 70% of human cases in France: SE (33%), ST (32%), and *S.* Hadar (6%) [7].

Among the *Salmonella* serotypes that were most isolated in Mexico between 1972 and 1999 were SE, ST, *S.* Derby, *S.* Agona, and *S.* Anatum, in decreasing order [8].

The clinical form of SE infection usually manifests as an episode of self-limiting enterocolitis, with symptoms that resolve within five days. It takes 8–72 hours for the infection to manifest itself, with clinical signs of diarrhea and intestinal pain. Antibiotics are not usually required in most cases of recovery. Although rare, severe diarrhea can occur, and a person may become ill to the point where they require hospitalization. Age (Children and elderly) and immunocompromised individuals are more susceptible than the general population. The infection in these patients can move from the intestines into the bloodstream and then to other organs, potentially leading to death unless the patient receives quick treatment with antibiotics [9].

Salmonella Enteritidis is a bacterium that causes intestinal infection in various animal species, particularly birds, without showing any symptoms. A strain of SE enters the ovaries of otherwise healthy hens, infecting the eggs before the shell is formed and contaminating the eggs, causing high mortality rates in neonate chickens [7].

Campylobacter jejuni is surpassing infections caused by *Salmonella* spp. and *Shigella* spp. in developed countries [10]. Most of the time, the origins of this infection are associated with animal feed, specifically poultry products [11]. In the United Kingdom and Denmark, outbreaks of campylobacteriosis are related to the consumption of undercook poultry products [12, 13]. Campylobacteriosis in animals destined for slaughter is rare in Mexico, and the disease's influence on human health is unknown.

2. Characterization of salmonellosis

Salmonella belongs to the Enterobacteriaceae family. They are Gram-negative bacilli that do not form spores. In this genus, there are three types of antigens: somatic O, flagellar H, and capsular Vi, which are used to distinguish more than 2500 serotypes based on their agglutination properties, which are used to determine more than 2500 serotypes. New serotypes are added to the Kauffmann-White list every year, which is updated with the latest information [14].

Salmonella is a genus that contains only two species: *salmonella bongori* and *S. enterica*, which is subdivided into six subspecies: *entericae, salamae, arizonae, diarizonae, houtenae,* and *indica*. Salmonellosis in humans and higher animals is caused by serotypes of the subspecies *entericae*, which account for nearly 99% of all cases [15]. A serotype of *S. enterica*, subspecies *entericae*, serotype *enteritidis*, abbreviated SE is used for practical diagnostic and epidemiological purposes [16].

Serotypes can be further separated by developing biotypes and phagotypes, which can then be used for more detailed studies of taxonomy and pathogenesis. The biotype denotes the biochemical variance between organisms of the same serotype, whereas the phenotype states the varied vulnerability of organisms of the same serotype to bacteriophage lysis [16].

Salmonella enterica serotypes Typhi and Parathyphi produce severe infections in humans known as a septicemic syndrome and typhoid fever, respectively, but are not pathogenic to animals. *S. Gallinarum* and *S. Abortus-ovis* cause avian typhoid and abortions in sheep, respectively, but rarely cause mild or asymptomatic infections in people. However, there are serotypes of *S. Choleraesuis* that produce severe disease in its usual carrier, the pig, but can also be pathogenic in humans. SE and ST infect both people and animals, however, in the latter, primarily hens, they cause asymptomatic illnesses [15, 17]. Undercook poultry products and the increased antibiotic resistance are linked to the increased number of infections in humans [7].

The mechanisms causing the rise in SE infections in birds have not been fully identified, making it challenging to identify illness in otherwise healthy chickens. Infections with SE in many animals, particularly chickens, with no apparent clinical indications and no acute outbreaks with mortality have been identified [16]. However, these healthy carriers can spread infection by fecal contamination of meat and egg. It is challenging to detect SE when the number of bacteria present is less than 9% [18].

The mechanization of poultry production and the export of parent birds have both contributed to the global spread of SE. For example, in the USA, molting of laying hens is a common practice that reduces or eliminates feeding of the birds' weight loss in birds; this practice speeds up molting but renders chickens more susceptible to SE infection, and once infected, they excrete the microorganism in feces in significantly high concentrations, which in turn increases the risk of egg contamination [19]. In poultry, vertical transmission to the progeny is common [20]. According to research in the Netherlands, flocks of laying hens, are primarily infected through direct contact with contaminated farm environments [21]; however, the epidemic that occurred in the United Kingdom in the early 1980s is attributed to the introduction of lines of progenitor birds infected with the phage type 4 [22].

S. Enteritidis can be introduced into flocks by rodents, which are highly vulnerable to infection, to the point that purposeful infection was employed to eliminate mice [23, 24]. *S. Enteritidis*, which was employed as a pesticide in the United Kingdom in 1940, was a type 6 phage [24, 25]. However, it has been demonstrated that acquiring the IncX plasmid changes phage 4 to phage 6 strain [26].

The dramatic increase in infection with *Salmonella Enteritidis*, a type 4 phage in humans in Europe since 1980, suggests that the bacterium has recently acquired new virulence genes [27].

More microbial genomes have been sequenced and compared recently, allowing the frequency of mutations to be approximated. The recombination mechanisms implicit in the replication process through the acquisition or loss of gene-carrying areas are a significant source of evolution. Plasmids, genomic islands, bacteriophages, transposons, and insertion sequences are other mechanisms of transferring or acquiring virulence genes [27]. These mobile components provide advantages to microbes in adapting to infecting specific cells [28]. Pathogenicity islands, or genes associated with virulence, arise outside of bacteria as mobile elements. Acquired pathogenicity islands contribute to the aggressive nature of bacteria by containing clusters of genes that boost virulence and can change a benign organism into a pathogenic one. Twelve islands of pathogenicity for *Salmonella* spp. have been described, some of which are shared by all serotypes of the species, while others are exclusive to individual serotypes [7].

Salmonella Gallinarum can induce flock immunity against serotype 09, indicating cross-immunity with *Salmonella Enteritidis*. As a result of this immunological feature, it has been proposed that to the extent that *Salmonella Gallinarum* has been removed by vaccination and slaughter of afflicted birds, its elimination may have allowed *Salmonella Enteritidis* to establish itself [7, 29]. Conversely, in Great Britain, the 50% decrease in *Salmonella Enteritidis* infection in birds since 1997 corresponds to the introduction of new live vaccines against serotype 09, in place of vaccines with bacteria killed in formalin [22]. For practical purposes, in terms of controlling transmissible zoonoses, vaccination of birds against *Salmonella Enteritidis* of serotype 09 may be indicated even in situations where *Salmonella Gallinarum* has been eradicated.

Integrons, which typically carry one or more antibiotic resistance genes, are another key source of microbial variety [30]. The growth and spread of antibiotic-resistant bacteria is an unavoidable side consequence of antibiotic treatment. Some *Salmonella* strains are resistant to most common antibiotics, including fluoroquinolones, with the latter resistance being related to point mutations in the gyrA gene [31].

Considering that integrons may spread antibiotic-resistant genes in addition to transposons, genomic islands, and plasmids, treatment with antimicrobial agents may contribute to the increase in the population of bacteria resistant to related antimicrobial agents, and therefore the use of antimicrobials in animal feed may have adverse effects on human health, for its selection effect on the resistant bacterial population [7].

3. Characterization of campylobacteriosis

The classification of the genus *Campylobacter* has been revised, and 16 species are now acknowledged [32]. These bacteria have spiral or curved rod shapes, are Gram-negative, have flagella that let them move, and are microaerophilic. The three species of medical and veterinary significance are as follows: *C. jejuni, Campylobacter coli,* and *C lari*. *C. jejuni* is divided into two subspecies: *C jejuni jejuni*, referred to simply as *C. jejuni*, which is associated with disease to humans, and *Campylobacter jejuni doylei*, which only sporadically affects humans [33].

Campylobacter spp. are oxidase-positive, reduce nitrates, are methyl red and Voges-Proskauer negative, and do not hydrolyze gelatin. Except for some strains of *Campylobacter lari*, most species are urea negative. Microorganisms that have been exposed to water for an extended period take the form of coconuts, which are more challenging to develop and may not even be cultivable. *C. jejuni*, *C. coli*, and *C. lari* are thermophilic, meaning they grow best at 42°C and 43°C and do not grow at temperatures lower than 25°C [34, 35]. Culture in selective media takes two days, and confirmatory testing on the species takes two more days [36].

Human infection is restricted to the gastrointestinal system and results in various forms of diarrhea. Infection can induce neurological abnormalities in rare cases [37]. The majority of illnesses are caused by the consumption of chicken and pork. In addition, *C. jejuni*, *C. coli*, and *C. lari* cause gastroenteritis in humans. Nevertheless, *C. lari* derived from pigs accounts for just 3% of the isolates [10].

4. Relevant aspects of pathogenicity mechanisms

Salmonella Enteritidis causes infection by attaching the intestinal mucosa and then invading the enterocytes. *Salmonella Enteritidis* adheres to the surface of enterocytes via the fimbriae and flagella. *Salmonella Enteritidis*' primary fimbriae are SEF14, SEF17, and SEF21. *Salmonella Enteritidis* colonization in birds occurs primarily in the cecum. Glycosphingolipid (GSL) GlcCer (N-1) and ganglioside GM3 (G-1) from chicken intestinal mucosa, found in ileum and cecum, have been studied as *Salmonella Enteritidis* SEF21 fimbria receptors [38].

When *Salmonella Enteritidis* crosses the epithelium and reaches the intestinal lamina propria, it invades the macrophages, and as it is generally resistant to the action of these, these cells serve as a vehicle to invade other organs [39].

C. jejuni and *C. coli* are commensals that reside in the intestines of numerous animals, including poultry, and survive in temperatures as low as 4°C for several weeks [40].

Campylobacter jejuni, produces enterotoxins and cytotoxins, which are the main cause of digestive symptoms in humans [33].

Until recently, the primary source of *Campylobacter* transmission in birds was thought to be horizontal from garbage, water, insects, equipment, and wildlife. Given the failures in attempts to grow *Campylobacter* from incubators or newborn chicks [41, 42]. However, considering that *C. jejuni* has been isolated in the reproductive organs of hens and the sperm of parent roosters, vertical transmission through the egg must be considered as a probable route of infection [43].

5. The problem of diagnosis and detection

The main problem is that *Salmonella* and *Campylobacter* are inhabitants of the intestine of birds, and in that environment, many bacterial organisms grow. Therefore, when trying to isolate a pathogenic species, it may not be possible to detect if the number is proportionally deficient and hidden by other organisms' growth. For this reason, the use of immunological and molecular biology techniques has been recommended to detect the existence of carrier animals in a short time. Isolation and identification of *Campylobacter* are problematic since it is slow-growing and easily confused with bacteria of the genus *Arcobacter*. The biggest drawback is that these are inert organisms that do not metabolize the sugars that are traditionally used to differentiate enterobacteria. That different environmental conditions, temperature, and antimicrobial sensitivity, as well as the hydrolysis of hippurate and indoxyl acetate, are used during isolation of *Campylobacter* [44].

Salmonella detection techniques based on polymerase chain reaction (PCR) have been developed for naturally and artificially contaminated food [45]. Several sensitive and precise PCR approaches for the detection of *Campylobacter spp*. are also available, which reduce diagnosis time to 48 hours and can detect up to one CFU (colony forming unit)/gram of sample [33, 46].

6. Foresight in Mexico

Unpublished results (isolation and identification) from a recent study in a commercial poultry company in Mexico revealed that from 30 broiler chickens, 8 chickens were positive for SE, and 2 were positive for ST. Without intending to conclude a single farm's sampling, given the homogeneous conditions under which modern poultry farming is carried out, these preliminary findings suggest that the prevalence of *Salmonella* in poultry farms in Mexico may be comparable to that of other countries with technical poultry farming. In the same study, 9/30 chickens were *C. jejuni* positives, and 2/30 were *C. coli* positives. According to these preliminary findings, *C. jejuni* and *C. coli* are likely to live as commensals in the intestinal tracts of broiler chicken in commercial flocks in Mexico, implying that studies involving representative segments of national poultry farming will be helpful soon. Campylobacteriosis in people must be expected in Mexico; hence, epidemiological research is a top priority. Given its sensitivity and specificity, the PCR approach is recommended to detect salmonellosis and campylobacteriosis in animals, food, and the environment to strengthen the foundations for the development of relevant epidemiological markers in Mexico.

Author details

Adriana del Carmen Gutiérrez-Castillo[1], Leopoldo Henri Paasch-Martínez[2] and Norma Leticia Calderón-Apodaca[2]*

1 Faculty of Veterinary Medicine and Zootechnics, Autonomous University of the State of Mexico, Toluca City, State of Mexico, Mexico

2 Faculty of Veterinary Medicine and Zootechnics Autonomous University of Mexico UNAM, Department of Avian Medicine, Mexico City, Mexico

*Address all correspondence to: nlca@unam.mx

References

[1] Hinton MH. Infections and intoxications associated with animal feed and forage which may present a hazard to human health. Veterinary Journal. 2000;**159**:124-138

[2] World Health Organization. Global aspects of emerging and potential zoonoses: A WHO perspective. Emerging Infectious Diseases. 1997;**3**:223-228

[3] International Food Safety Authorities Network (Infosan) Red Internacional de Autoridades de Inocuidad de los alimentos. Nota de información INFOSAN 3/2005-*Salmonella*. 13 de abril de 2005

[4] European Commission. 2004. Trends and sources of zoonotic agents in animals, feeding stuffs, food and man in the European Union and Norway in 2002. European Commission, health and consumer protection directorate-general. Sanco/29/2004. http://europa.eu.int/comm/food/food/biosafety/salmonella/zoonoses reps 2002 en.htm

[5] Kapperud G, Stenwig H, Lassen J. Epidemiology of *Salmonella typhimurium* 0:4-1 infection in Norway: Evidence of transmission from an avian wildlife reservoir. American Journal of Epidemiology. 1998;**147**:774-782

[6] Wierup M, Engström B, Engavall A, Wahlström H. Control of *Salmonella enteritidis* in Sweden. Food Microbiology. 1995;**25**:219-226

[7] Velge P, Cloeckaert A, Barrow P. Emergence of *Salmonella* epidemics: The problems related to *Salmonella* enterica serotype Enteritidis in multiple antibiotic resistance in other major serotypes. Veterinary Research. 2005;**36**:267-288

[8] Gutiérrez-Gogco L, Montiel-Vázquez E, Aguilera-Pérez P, González-Andrade MC. Serotipos de *Salmonella* identificados en los servicios de salud de México. Salud Pública de México. 2000;**42**:490-495

[9] Food and Agriculture Organization of the United Nations/World Health Organization. Discussion Paper on Risk Management Strategies for *Salmonella* spp in Poultry. Joint FAO/WHO Food Standards Programme Codex Committee on Food Hygiene. Orlando, USA: FAO; 2003. p. 11

[10] Lamoureux M, Mackay A, Messier S, Fliss I, Blais BW, Holley RA, et al. Detection of *Campylobacter jejuni* in food and poultry viscera using immunomagnetic separations and microtitre hibridation. Journal of Applied Microbiology. 1997;**83**:641-651

[11] Nielsen EM, Nielsen NL. Serotypes and typability of *Campylobacter jejuni* and *Campylobacter coli* isolated from poultry products. International Journal of Food Microbiology. 1999;**46**:199-205

[12] Pearson AD, Greenwood MH, Feltham RK, Healing TD, Donaldson J, Jones DM, et al. Microbial ecology of *Campylobacter jejuni* in a United Kingdom chicken supply chain: Intermittent common source, vertical transmission, and amplification by flock propagation. Applied and Environmental Microbiology. 1996;**62**:4614-4620

[13] Neimann J, Engberg J, Molbak K, Wegener HC. Risk Factors associated with sporadic campylobacteriosis in Denmark. 4th World Congress, Food borne Infections and intoxications. Berlin, Germany; Federal Institute for Risk Assessment. FAO/WHO Collaborating Centre for Research and Training in Food Hygiene and Zoonoses. 1998

[14] Popoff MY, Bockemuhl J, Gheesling LL. Supplement 2001

(No. 45) to the Kauffmann-White scheme. Research in Microbiology. 2003;**154**:173-174

[15] Uzzau S, Brown DJ, Wallis T, Rubino S, Leori G, Bernard S, et al. Host adapted serotypes of *Salmonella* enterica. Epidemiology and Infection. 2000;**125**:229-255

[16] Ward LR, de Sa JD, Rowe B. A phagetyping scheme for *Salmonella enteritidis*. Epidemiology and Infection. 1987;**99**:291-294

[17] Duchet SM, Mompart F, Berthelot F, Beaumont C, Lechopier P, Pardon P. Differences in frequency, level and duration of cecal carriage between four outbreed chicken lines infected orally with *Salmonella enteritidis*. Avian Diseases. 1997;**41**:559-567

[18] Humphrey TJ. Contamination of egg shell and contents with *Salmonella enteritidis*: A review. International Journal of Food Microbiolology. 1994;**21**:31-40

[19] Holt PS. Molting and *Salmonella* enterica serovar Enteritidis infection: The problem and some solutions. Poultry Science. 2003;**82**:1008-1010

[20] Mollenhorst H, Van Woudenbergh CJ, Bokkers EGM, y Boer IJM. Risk factors for *Salmonella enteritidis* infection in laying hens. Poultry Science. 2005;**84**:1308-1313

[21] Van de Giessen AW, Ament AJHA, Notermans SHW. Intervention strategies for *Salmonella enteritidis* in poultry flocks: A basic approach. International Journal of Food Microbiology. 1994;**21**: 145-154

[22] Ward LR, Therefall J, Smith HR, O'Brien SJ, Riemann H, Kass P, et al. *Salmonella enteritidis* epidemic. Science. 2000;**287**:1753-1754

[23] Henzler DJ, Opitz HM. The role of mice in the epizootiology of *Salmonella enteritidis* infection on chicken layer farms. Avian Diseases. 1992;**36**:625-631

[24] Friedman CR, Malcom G, Rigau-Perez JG, Arambulo P, Tauxe RV. Public health risk from *Salmonella*-based rodenticides. Lancet. 1996;**347**:1705-1706

[25] Guard PJ, Henzler DJ, Rahmann MM, Carlson RW. On-farm monitoring of mouse-invasive *Salmonella* enterica serovar Enteritidis and a model for its association with the production of contaminated eggs. Applied and Environmental Microbiology. 1997;**63**:1588-1593

[26] Threlfall EJ, Ridley AM, Ward LR, Rowe B. Assessment of health risk from *Salmonella*-based rodenticides. Lancet. 1996;**348**:616-617

[27] Helmuth R, Schroeter A. Molecular typing methods for *S.enteritidis*. International Journal of Food Microbiology. Jan 1994;**21**(1-2):69-77

[28] Ochman H, Lawrence JG, Groisman EA. Lateral gene transfer and the nature of bacterial innovation. Nature. 2000;**405**:299-304

[29] Chacana PA, Terzolo HR. Protection conferred by a live *Salmonella* enteritidis vaccine agaist fowl typhoid in laying hens. Avian Diseases. 2006;**50**:280-283

[30] Boyd D, Peters GA, Cloeckaert A, Boumedine KS, Chaslus-Dancla E, Imbereschts H, Mulvey MR. Complete nucleotide sequence of a 43-kilobase genomic island associated with the multidrug resistance region of Salmonella enterica serovar Typhimurium DT104 and its identification in phage type DT120 and serovar Agona. Journal of Bacteriology. Oct 2001;**183**(19):5725-5732

[31] Angulo FJ, Johnson KR, Tauxe RV. Origins and consequences of antimicrobial-resistant nontyphoidal

Salmonella: Implications for the use of fluoroquinolones in food animals. Microbial Drug Resistance. 2000;**6**:77-83

[32] Petersen L, Nielsen EM, On SWL. Serotype and genotype diversity and hatchery transmission of *Campylobacter jejuni* in commercial poultry flocks. Veterinary Microbiology. 2001;**82**:141-145

[33] Phillips CA. Incidence, epidemiology and prevention of food borne *Campylobacter* species. Trends in Food Science and Technology. 1995;**6**:83-87

[34] Federighi M. *Campylobacter* et hygiène des aliments. Paris, France: Ed. Polytechnica; 1999

[35] Talibart R, Denis M, Castillo A, Cappelier JM, Ermel G. Survival and recovery of viable but noncultivable forms of *Campylobacter* in aqueous microcosm. International Journal of Food Microbiology. 2000;**55**:263-267

[36] Wassenaar TM, Newell DG. The Genus Campylobacter. 2006. January 2006. DOI: 10.1007/0-387-30747-8_4. In book: The Prokaryotes. Available from: https://www.researchgate.net/publication/279614401_The_Genus_Campylobacter

[37] Barrow PA, Page K. Inhibition of colonization of the alimentary tract in young chickens with *Campylobacter jejuni* by pre-colonization with strains of *C.jejuni*. FEMS Microbiology Letters. 2000;**182**:87-91

[38] Kramer J, Visscher AH, Wagenaar JA, Jeurissen SHM. Entry and survival of *Salmonella* enterica serotype Enteritidis PT4 in chicken macrophage and lymphocyte cell lines. Veterinary Microbiology. 2003;**91**:147-155

[39] Asheg AA, Levkut M, Revajova V, Sevcikova Z, Kolodzieyski L, Pistl J. Dynamics of lymphocyte subpopulations in inmmune organs of chickens infected with *Salmonella enteritidis*. Acta Veterinaria. 2003;**72**: 359-364

[40] Brooks BW, Robertson RH, García MM HD. Production and Western blot characterization of monoclonal antibodies specific for *Campylobacter jejuni* and *Campylobacter coli*. Journal of Rapid Methods and Automation in Microbiology. 1995;**4**:155-164

[41] Neil SD, Campell NJ, Greene JA. *Campylobacter* species in broiler chickens. Avian Pathology. 1984;**13**: 777-785

[42] Clark AG, Bueschkens DH. Survival and growth of *Campylobacter jejuni* in egg yolk and albumen. Journal of Food Protection. 1986;**49**:135-141

[43] Cox NA, Hofacre CL, Bailey JS, Buhr RJ, Wilson JL, Hiett KL. Presence of *Campylobacter jejuni* in various organs one hour, one day, and one week following oral or intracoacal inoculations of broiler chicks. Avian Diseases. 2005;**49**:155-158

[44] Oyarzabal O, Murphy B. *Campylobacter* and *Arcobacter*. International Food Hygiene. 1998;**9**:29-30

[45] Bülte M, Jakob P. The use of PCR-generated invA probe for the detection of *Salmonella* spp in artificially and naturally contaminated foods. International Journal of Food Microbiology. 1995;**26**:335-344

[46] Skanseng B, Kaldhusdal M, Knut R. Comparison of chicken gut colonizations by pathogens *Campylobacter jejuni* and *Clostridium perfringens* by real-time quantitative PCR. Molecular and Cellular Probes. 2006;**20**:269-279

Chapter 3

The Role of Immune Response and Microbiota on Campylobacteriosis

Ying Fu, Tahrir Alenezi, Ayidh Almansour, Hong Wang, Zhenquan Jia and Xiaolun Sun

Abstract

Million cases of campylobacteriosis and complications of post-*Campylobacter jejuni* infection occur every year around the world with huge life losses and economic burdens of billions of dollars. Few therapy options, such as antibiotics, are available to relieve severe cases of the enteritis. The slow progression on new intervention discovery and application is partially resulted from limited mechanistic understanding on campylobacteriosis pathogenesis. As a type of intestinal disorders, campylobacteriosis shares many common features with other intestinal diseases such as inflammatory bowel diseases (IBD) and *Clostridium difficile* infection. In pace with the advancement of the gastroenterology field, a large body of knowledge is accumulating on the factors influencing campylobacteriosis onset, development, and outcomes, including host immune response, intestinal microbiota, and its metabolites. In this chapter, we review the intestinal immune system, intestinal microbiome, and microbiome modulation of inflammation in the development of campylobacteriosis. The interplay between immunity, microbiota, and its metabolites may play essential roles on campylobacteriosis pathogenesis and the finding on the interaction may lead to new prevention and treatment options. The purpose of this chapter is to provide updated knowledge on the role of host–microbe interaction and the therapeutic potential on campylobacteriosis.

Keywords: colitis, infection, adaptive immunity, innate immunity, microbial metabolite, bile acids

1. Introduction

Campylobacter enteritis (also known as campylobacteriosis) is defined as an infection of the intestines that is manifested in the form of acute diarrhea followed by pain in abdomen, fever as well as other constitutional clinical indications [1]. Campylobacteriosis is a common foodborne pathogen disease worldwide caused by *Campylobacter jejuni* [2]. *C. jejuni* is a Gram-negative, microaerobic bacterium. Because of the large consumption and industrialized production of animal meat, the main reservoir of *C. jejuni* is food animals such as chickens and turkeys. *Campylobacter* is one of the most frequent causes of foodborne bacterial pathogen, particularly in developed countries. *C. jejuni* and *C. coli* are the foremost causes of infections in the vast majority of population [3]. According to CDC's report, 24% raw chicken meat carried *C. jejuni* [4]. Around 1.5 million cases reported in USA every year [5] and causing $6.9 billion losses annually [6].

C. jejuni is able to establish infection in the intestine with ingestion of minimum 500 viable bacteria, but the infection efficiency is influenced by host antibacterial defenses such as gut immune system and the intestinal commensal microbes [7]. The innate and adaptive immunity in gut actively surveils the luminal microbes, processes the intestinal cues, and establishes defense actions, resulting in constant gastrointestinal homeostasis [8]. The complex gut resident microbes live on and inside the host including bacteria, fungi, protozoa, viruses, and their metabolic products [9]. The gut microbes are important participants for food digestion, fermentation, and energy accommodation of intestinal tract [10]. During physiological process, metabolites, such as short chain fatty acid, bile acid, vitamins, and amino acids, are produced. The microbes, along with their metabolites play important roles in keeping the homeostasis of gastrointestinal immunity, and affecting their resistance to the invasion of pathogens [11]. In the following sections, we will have a detailed discussion on gut immunity, resident microbes, and their role on campylobacteriosis.

2. Intestinal immunity on campylobacteriosis

The immune system is comprised of a complex network of biological molecules and activities in organs, tissues, and cells to protect an organism against foreign substances or microbes (**Figure 1**). The immunity is generally categorized into two subsystems of innate and adaptive immunity [12]. The innate immunity initiates a quick immune response [13], while the adaptive immunity generates a comprehensive and long-lasting immune defense [12]. These two immune branches work closely together to defense host against the encountered foreign substances or microbes. The intestinal immunity is highly involved with *C. jejuni*-induced colitis. *C. jejuni*–induced severe campylobacteriosis in *Il10*$^{-/-}$ mice as showed by extensive intestinal immune cell infiltration, epithelial damage, goblet cell depletion and crypt hyperplasia and abscesses compared with uninfected mice [14]. In this section, we will briefly review recent advancement of intestinal immunity and campylobacteriosis.

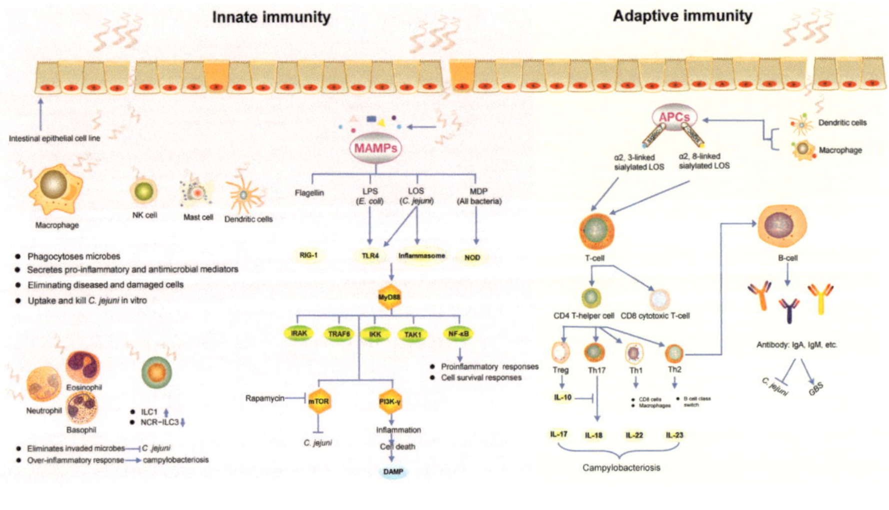

Figure 1.
Schematic illustration of the role of innate (left side) and adaptive (right side) immunity in campylobacteriosis.

2.1 Intestinal immunity and campylobacteriosis

In the gastrointestinal tract, the innate immunity is consisted of innate cells and soluble molecules, which are an important defense mechanism against foreign substances or microbes. The cellular innate immunity is consisted of various types of cells, including intestinal epithelial cell (IEC), granulocyte (neutrophil, basophil, and eosinophil,) dendritic cell (DC), macrophage, natural killer cell (NK), master cell, and innate lymphoid cell (ILC), and γδ T cell [15]. Only a single layer of IEC separates nearly sterile internal intestinal tissue from microbe-rich intestinal lumen, hence the integrity of IEC is essential for intestinal health. Notably, IEC line breakdown is often implicated in various intestinal disorders such as IBD [16], irritable bowel syndrome (IBS) [17], colorectal cancer [18], and *C. difficile* infection [19]. The destruction of IEC line and tissue upon invasion of *C. jejuni* [20] clearly demonstrates the important role of the epithelial cells. The innate effector cell of scavenging macrophage phagocytoses microbes and secretes both pro-inflammatory and antimicrobial mediators [21]. In addition, macrophage is essential for eliminating diseased and damaged cells through its programmed cell death. Macrophage uptakes and kills *C. jejuni in vitro* [22], although the role of macrophage in campylobacteriosis remains to be determined. Neutrophil is the most abundant type of granulocytes and consists of 40% to 70% of all white blood cells in humans [23]. Although neutrophil eliminates invaded microbes, the over-inflammatory response of neutrophil is responsible for the campylobacteriosis in a *Il10*$^{-/-}$ mouse model [24]. *C. jejuni*-induced colitis increases ILC1 (50% vs. 18%) but decreases NCR – ILC3 (13% vs. 43%) in the colonic lamina propria of germ free *Il10*$^{-/-}$ mice, compared to uninfected mice [25]. Effort is needed to investigate the role of various innate cells on campylobacteriosis pathogenesis.

At the molecular level, the innate cells recognize the microbes of their microbial-associated molecular patterns (MAMPs), such as lipopolysaccharides (LPS) and flagellin. MAMP is a component of a microbe and is sensed by innate cellular pathogen recognition receptors (PRRs), such as toll-like receptors (TLRs), nucleotide oligomerization domain (NOD) like receptors (NLRs), and retinoic acid inducible gene-I (RIG-I) like receptors (RLRs) [26, 27]. Previous articles have comprehensively reviewed the interaction between MAMP and PRR [28], hence we will not devote too much on them. Relevant to the topics of this chapter, LPS is expressed on the surface of Gram-negative bacteria such as *E. coli*, and it is recognized by TLR4 at the innate cell surface. *C. jejuni* expresses lipo-oligosaccharide (LOS) instead of LPS [29] and LOS is possibly recognized by TLR4 in DC and *Il10*$^{-/-}$ mouse model [20, 30]. Muramyl dipeptide (MDP) is the minimal bioactive peptidoglycan motif common to all bacteria and is sensed by NOD2 in innate cytoplasm. Microbiota-disturbed *Il10*$^{-/-}$; *Nod2*$^{-/-}$ mice are susceptible to *C. jejuni* induced colitis compared to *Il10*$^{-/-}$ mice [31], suggesting the role of NOD2 in host shown preventive mechanism against the pathogen. It is much needed to investigate various PRRs on detecting *C. jejuni* infection and to elicit immune response.

After trigged by PRRs detecting MAMP, innate response of a network of signaling pathways are activated, including TLR-MyD88/TRIF and inflammasome. MyD88 is a downstream adaptor protein of TLR and is essential for the signal transduction of the TLR signaling pathway [32]. The TLR signaling pathway is classified into either MyD88-dependent or MyD88-independent. With the exception of TLR3, all downstream signaling pathways of TLRs mediate through MyD88 [33]. For MyD88 dependent pathway, TLR signaling recruits and activates a number of molecules, including IRAK, TRAF6, TAK1, IKK, and NF-κB [32]. The TLR/Myd88/NF-κB signaling pathway then induces proinflammatory and cell survival responses. NF-κB signaling is activated in *C. jejuni*-induced colitis using germ free

Il10$^{-/-}$; NF-κB^{EGFP} mouse model. mTOR signaling is a downstream target of MyD88 and mediates *C jejuni*-induced colitis in *Il10*$^{-/-}$ and *Il10*$^{-/-}$; *Rag2*$^{-/-}$ mice, suggesting independence of T-cell activation [14, 24]. Blocking mTOR signaling with pharmacological inhibitor rapamycin attenuates *C jejuni*-induced intestinal inflammation, immune cell infiltration and the pathogen invasion, while rapamycin increases splenocyte autophagy [14]. In addition, *C. jejuni*–induced MyD88 downstream target PI3K-γ signaling mediates intestinal inflammation in *Il10*$^{-/-}$ mice through modulating neutrophil migration/infiltration into intestinal lamina propria [24]. During inflammation, damaged or dying cells release endogenous danger molecules called damage-associated molecular pattern (DAMP) such as high-mobility group box 1 (HMGB1), S100 proteins, and heat shock proteins (HSPs). The DAMP is sensed by TLR and inflammasome and is investigated extensively in non-infectious inflammation disorders [34]. Inflammasome is responsible for processing proIL1β and proIL18 into active forms [35]. It would not be surprised to find that DAMP-induced inflammation in *C. jejuni*-induced colitis, hence such work would yield important leads to understanding campylobacteriosis pathogenesis.

2.2 Intestinal adaptive immunity and campylobacteriosis

Despite the effective, fast, and general/non-specific response of innate immunity against infection, adaptive immunity is often developed in vertebrate animals, particularly in the case of unresolved innate response. With the assistance of innate immunity, the adaptive immunity of lymphocytes recognize and remember a foreign substance's or pathogen's unique antigens and builds an antigen-specific response to eliminate it [12]. Two major lineages of T and B lymphocytes are generated in the thymus and the bone marrow or the avian bursa of Fabricius [36]. The adaptive immunity mounts two types of activities: B cell mediated antibody responses, and T cell mediated immune response. DC, B-cell, and macrophage express specific "co-stimulatory" ligands recognized by co-stimulatory receptors on T cells, and are named antigen-presenting cells (APCs) for T cell activation. During the early developmental stages, B lymphocyte progenitor cells make somatic hypermutation for specific antibody, while T and B cells rearrange different sets of immunoglobulin (Ig) variable (V), diversity (D), and joining (J) gene segments to make the antigen binding regions of the T cell receptors (TCRs) and B cell receptors (BCRs) [37]. *Campylobacter* infection-induced Guillain-Barré syndrome (GBS), an autoimmune disease, demonstrates the implication of adaptive immunity in the pathogen infection.

T cells are grouped into two types based on the surface antigens: CD4-expressing T-helper cells, and CD8-expressing cytotoxic T-cells [38]. It remains elusive the role of CD8 cells in campylobacteriosis, but accumulating evidence supports the notion on the important role of CD4 cells in campylobacteriosis pathogenesis. The major intestinal CD4^{+} T cells are T help cell 1 (Th1), Th17, and regulatory T cell (Treg, Foxp3-expressing) cells, although Th2, Th9, Th22, follicular helper T (Tfh), iTreg, and type 1 regulatory T cell (Tr1) are present [39, 40]. The adaptive immunity is actively influenced by innate immunity. In gut lamina propria, intestinal innate tolerogenic CD103^{+} DCs induce FoxP3+ Tregs by stimulating CCR7 and integrin-$\alpha_{IV}\beta_7$ on T cells resided in mesenteric lymph nodes [41–43]. The differential interaction between *Campylobacter* LOS and siglec-7 receptors on APC cell-surface influences the fate of naïve CD4 cells into different type of effector Th cells [44, 45]. Specifically, siglec-7 receptors on APC binds with α2, 8-linked sialylated LOS induces the Th1 polarization, while its interaction with α2, 3-linked sialic acid induces a Th2 development [45]. Generally, Th1 cells activate more cytotoxic CD8 cells and macrophages to enhance immunity against the invading or intracellular microbes, while Th2 cells mediate class switching of B-cells to eliminate the extra-cellular microbes [38]. Besides Th1 and Th2 cells, Th17, Th22, and Treg may also be induced in campylobacteriosis as evidenced by the elevated cytokine markers of IL-17, IL-18, IL-22, IL-23, and IL-10 in patients' serum following infection with *Campylobacter* [26, 46].

After Th2 cell activation, B-cells are induced to produce antibody (Ab) against *Campylobacter* infection. At 7 days post-infection (acute phase), blood Abs, IgA and IgM increase in serum [47]. From 1 week up to 1-year post-infection (convalescent phase), anti-*Campylobacter* Ab is detectable in serum and saliva of campylobacteriosis patients and could protect the subjects against subsequent *Campylobacter* infection [48, 49]. Similarly, IgA and IgM are persistent in campylobacteriosis patients for up to 20 days or 2 months post-infection. The downside to the adaptive humoral response is the incidence of GBS. *Campylobacter* often alters its LOS outer core to mimic human neuronal gangliosides for escaping from the host immune system but resulting in GBS [50]. α2, 3-linked sialic acid in *C. jejuni* LOS is one of the culprits. Developing effective vaccine or monoclonal antibody to control *C. jejuni* infection is, therefore, imminent.

3. Intestinal microbiome and campylobacteriosis

Human body, particularly gastrointestinal tract, inhabits trillions of diverse microbes including bacteria, archaea, virus, and eukarya [51]. These microbes called microbiota (**Figure 2**), and their metabolic activities and metabolites are collectively named microbiome [52]. The microbiota demonstrates a complex and diverse phylogeny of notable microbial species [53–55]. The human microbiota is comprised of 2172 prokaryotic species and the main phyla are Bacteroidetes, Firmicutes, Proteobacteria, and Actinobacteria [56]. The inhabitant gut microbiota influences important biological processes, such as metabolism of food, production of fat and vitamins, activation of angiogenesis as well as safeguard against adversary pathogens [53, 54]. Relevant to the topic of this chapter, the colonization of gut microbiota effectively inhibits the colonization and excessive growth of

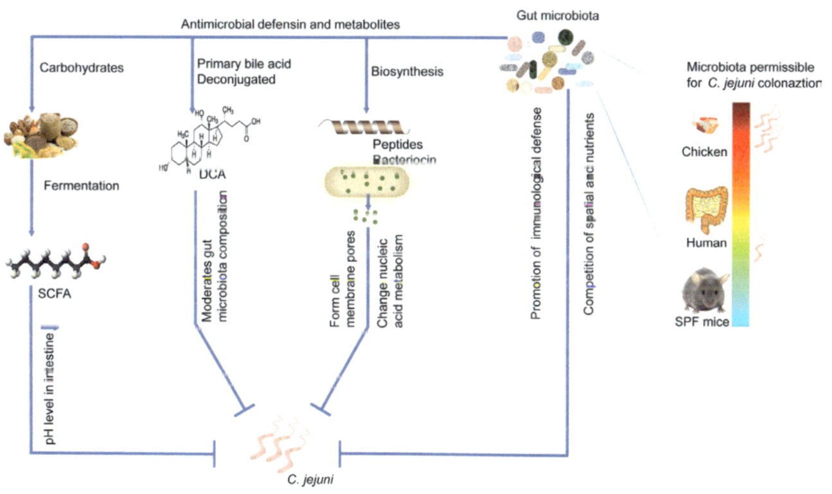

Figure 2.
Schematic illustration of the role of microbiota and its metabolites in campylobacteriosis.

potential pathogenic microbes, called colonization resistance [55]. The colonization resistance is through various mechanisms including direct competition of spatial and nutrients, production of antimicrobial defensin and metabolites, and indirect inhibition via stimulation of innate and adaptive immunity [57]. Certain microbiota phyla reduction is associated with an abolished biological colonization resistance [7, 58]. The colonization resistance, therefore, prevents pathogen attachment to the respective target site, depletes nutrients, and blocks virulence expression.

During its metabolism of nutrients, microbiota synthesizes varied range of metabolites and related small molecules [59–62]. It is recognized that the microbiota metabolites are absorbed across the gastrointestinal tract in circulation and impact host physiology [63–65]. Accumulating findings strongly support the important role of microbiota metabolites against gut pathogens. One example of the metabolite is short-chain fatty acids (SCFA). SCFA is fermented from carbohydrates (e.g., starch and fiber) and influences the gut microbiota community by reducing luminal pH level [66–69]. Another abundant microbiota metabolite is bile acid. Bile acids produced in the liver are excreted into the intestine as conjugated (taurine or glycine) forms to facilitate in digestion of dietary lipids. The bile acids are deconjugated in small intestine by bile salt hydrolases (BSH) [70] and absorbed up to 95% along intestinal line through enterohepatic cycle [71]. Furthermore, microbiota produces bacterial toxic and short peptides (e.g. bacteriocin) and bacterial toxins to inhibit the growth and colonization of other species [72]. The bacterial toxic peptides are categorized into those produced by Gram-negative bacteria (mostly by *Enterobacteriaceae*) and those produced by Gram-positive bacteria (lactic acid bacteria and some *Streptococcus* species) [73, 74]. The peptides are further classified into subgroups based on molecular weight, such as microcins (lower molecular weight peptides) and colicins (higher molecular weight proteins). The inhibition mechanism of bacteriocin is to change nucleic acid metabolism and to form cell membrane pores for eliminating other bacteria [75–78]. In this section, we will briefly review recent advancement on the interaction of microbiome and campylobacteriosis.

3.1 Microbiota and campylobacteriosis

To colonize in the gut, *C. jejuni* has to overcome numerous hurdles and endures in diverse environments. With minimum 500 viable bacteria, the pathogen has to establish in the intestine against host antibacterial defenses such as the intestinal bile acids and the intestinal microbiota [7]. The pathogenesis of *Campylobacter*-induced enteritis remains elusive because of lacking reliable animal models. Notably, *C. jejuni* is often colonized in birds without any pathological symptom [79, 80], while specific pathogen free mice, but not germ-free mice, are resistant to the pathogen colonization [24, 81]. Humans are susceptible to *C. jejuni*-induced enteritis, but the pathogen is often cleared within 1 to 2 weeks [82]. The reason why *C. jejuni* colonizes animals differentially remains elusive. Because the intestinal microbiota is different between animals, it is possible the gut microbiota influences bacterial pathogen colonization [55].

Chickens are susceptible to *C. jejuni* asymptomatic colonization and their microbiota could be friendly to *C. jejuni* infection. In 35-day old broiler chickens, families *Lactobacillaceae* and *Clostridiaceae* in the ileum and *Lachnospiraceae* and *Ruminococcaceae* in ceca are dominant, while genera *Ruminococcus* and *Oscillospira* account for 35% in ceca operational taxonomic units (OTUs) [80]. In a field study of 35-day old broiler chickens at four farms in Italy, the relative abundance of class *Clostridiales* is higher in caeca of *Campylobacter*-negative farms than positive farms, while *Bacteroidales* is the opposite (80.0% vs. 65.7%) [83]. In 56-day old broiler chickens, *C. jejuni* colonization is associated with reduced genera abundance of *Corynebacterium* and *Lactobacillus* but increased genera *Ruminococcaceae* and *Streptococcus* [84]. The authors also found that *C. jejuni* colonization is positively associated with genera *Escherichia*, *Alistipes*, *Enterococcus*, *Bacteroides*, *Shigella*, *Gallibacterium*, *Campylobacter*, *Faecalibacterium*, *Blautia*, *Enterobacter* and *Clostridium*. In mice, two genera of *Clostridium sensu stricto* and *Enterococcus* are associated with mice susceptible to *C. jejuni*-induced colitis in a microbiota transplantation model [85]. Ampicillin treatment increases cecal genus *Barnesiella* but reduced *Clostridium XIVa* in the microbiota of *C. jejuni*-susceptible mice [86]. The abundance of *E. coli* is positive associated with *C. jejuni* colonization in mice [87]. Human campylobacteriosis patients have an increased abundance of genera *Escherichia*, *Bacteroidetes*, *Phascolarctobacterium*, and *Streptococcus* in stool [3]. Comparably, microbiota sequencing data from cross-sectional IBD patients showed that IBD is associated with dysbiosis characterizing by reduced gut bacterial diversity, together with increased genera *Fusobacterium*, *Escherichia, Faecalibacterium*, *Roseburia, Ruminococcaceae*, *Peptostreptococcaceae, Christensenellaceae*, and *Collinsella* [88]. The changes in the gut microbiota of IBD patients show an increase in facultative anaerobes, including *Escherichia coli* [89], and a decrease in obligately anaerobic [90]. IBD patients with active disease have increased gut *Enterococcus, Fusobacterium, Haemophilus, Megasphaera, Campylobacter*, while *Roseburia, Christensenellaceae, Oscillibacter, and Odoribacter* are enriched in the gut of IBD patients with inactive disease [88]. Although increasing evidence supports the role of microbiota promoting *C. jejuni* infection or other enteritis, additional studies are much needed.

On the other hand, SPF mice are resistant to *C. jejuni* colonization and their microbiota could be hostile to *C. jejuni* colonization. Through mining 16S DNA sequencing datasets, the core microbiota of healthy mice in cecum is found to be comprised of 37 genera, including *Anaerostipes, Parabacteroides, Anaerotruncus, Oscillibacter, Clostridium XlVb, Flavonifractor, Bacteroides, Barnesiella, Alistipes, Helicobacter, Saccharibacteria, Prevotella, Lachnoanaerobaculum, Lactobacillus, Intestinimonas, Roseburia, Alloprevotella, Rikenella, Enterorhabdus, Erysipelotric haceae_incertae_sedis, Eggerthella, Allobaculum, Lachnospiracea_incertae_sedis, Pseudoflavonifractor, Bifidobacterium, Marvinbryantia, Mucispirillum, Clostridium XIVa, Blautia, Anaerofilum, Parasutterella, Odoribacter, Olsenella, Turicibacter, Gordonibacter, Ruminococcus,* and *Acetatifactor* [91]. Eight genera of *Clostridium XI, Oscillibacter, Bifidobacterium, Butyricicoccus, Hydrogenoanaerobacterium, Lactobacillus, Roseburia,* and *Coprobacillus* are increased in the microbiota of *C. jejuni*-resistant mice [85]. Supplementation of *Bifidobacteria* and *Lactobacillus* species has been shown to reduce the colonization of *Campylobacter* in birds [92–95]. The probiotics against *Campylobacter* colonization are through promotion of immunological defense mechanisms such as stimulation of defensins and interleukins as well as alteration of integrity of epithelial cell barrier [96].

In human subjects with *Campylobacter*-negative, the abundance of genera *Clostridiales*, unclassified *Lachnospiraceae*, and *Anaerovorax* are increased [3]. Comparably, the *Campylobacter*-negative individuals showed increased abundance of family *Lachnospiraceae,* particularly its two genera *Dorea* and *Coprococcus* [97]. People who consume plant-based low fat and polysaccharide rich diet are more resistant to *C. jejuni* infection compared to individuals consuming western diet [98]. Hence, increasing studies are being performed to investigate the role of microbiota against *C. jejuni* infection.

3.2 Microbial metabolites and campylobacteriosis

The questions following section 3.2 are how microbiota facilitates or reduces *C. jejuni* infection. Besides direct inhibition by competition of space and nutrients [53, 54], microbiota metabolites may exert indirect antagonism against *C. jejuni*. The intestinal microbiota generates a variety of bioactive metabolites after metabolizing nutrients from diets and host secretions. A few data are available on the relationship of microbiota metabolites and *C. jejuni* infection, but accumulating data are present in the field of IBD (Crohn's Disease-CD and Ulcerative Colitis-UC), a close enteritis to campylobacteriosis. The metabolomics of IBD patients is shifted from healthy subjects with characterization of increased bile acids, taurine, and tryptophan [99]. Out of the 2,729 differentially abundant metabolites, the majority (71%) are significantly depleted in IBD relative to non-IBD controls; 8% are significantly elevated in both CD and UC; 19% are specifically elevated in CD; and only 3% are specifically elevated in UC [100]. Specifically, IBD enriches lactate, sphingolipids, and primary bile acids of cholate (CA) and chenodeoxycholate (CDCA) but with reduction of triterpenoids, pantothenate, long-chain fatty acids, phenylbenzodioxanes, cholesterols (including cholestenone), triacylglycerols (TAGs), and secondary bile acids deoxycholic acid (DCA) and lithocholic acid (LCA). Interestingly, IBD patients show a decrease in obligately anaerobic producers of short-chain fatty acids [90].

Furthermore, IBD patients have increased polyunsaturated fatty acids (e.g., adrenate and arachidonate) but reduced pantothenate and nicotinate [101]. CD patients have increased levels of conjugated and sulfated bile acids in the feces [102]. In a functional analysis with shotgun metagenomics data, sulfur metabolism is identified with an enrichment of sulfonate, methionine, cysteine and taurine transport systems in mice colonized with microbiota from active IBD patients [88]. These metabolic changes are consistent with the increased abundance of sulfate-reducing bacteria (e.g., *Desulfovibrio*, *Clostridia, Bilophila*, and *Bacteroides fragilis*), some of which use sulfate as a terminal electron acceptor for respiration and concomitantly produce hydrogen disulfide, a toxic metabolic byproduct [88]. In accordance with IBD, secondary bile acid DCA, but not LCA and ursodeoxycholic acid, reduces *C. jejuni* counts and moderates intestinal microbiota composition in broiler birds [79]. DCA also reduces *C. jejuni*-induced colitis in ex-germ free *Il10*$^{-/-}$ mice [85]. Together, microbiota metabolites play an essential role on enteritis such as campylobacteriosis and IBD, and finding additional metabolites will assist development of therapeutic agents.

One specific and well-studied bacteria-bacteria interaction through microbial metabolites is called quorum sensing (QS) [103]. When the number of bacteria in the surrounding environment reaches certain level, bacteria activate QS and release specific signaling molecules of autoinducers (AIs) to modulate the expression of themselves and surrounding others on virulence, the ability for invasion and colonization, and the formation of biofilm [104]. Two types of AIs have been studied. AI-1 is produced by N-acyl-homoserine lactones (AHL) synthase and mediates intraspecies communication in Gram-negative bacteria. AI-2 is produced by S-ribosylhomocysteine lyase (LuxS) and mediates both intra- and interspecies communication in Gram-positive or Gram-negative bacteria [103, 104]. LuxS/AI-2 system plays important roles in cell–cell interactions in *C. jejuni* [105, 106]. Because biofilm is crucial for *C. jejuni* survive outside of hosts and facilitates its transmission from chicken reservoirs to humans, LuxS-mutant strains show deficient in biofilm formation and possible reduction of their transmission [107]. *C. jejuni* 81176 luxS mutant shows significant decreased colonization in chickens [108]. Deletion of luxS gene in *C. jejuni* NCTC IA3902 strain completely inactivates its colonization in the intestinal of chickens [109] or guinea pig [110], while the complemented strain with luxS gene restores the colonization ability comparable to the wildtype. It remains largely elusive what is the role of QS in the interaction of microbiota and *C. jejuni* and on the pathogen infectious capacity of colonization and induction of intestinal inflammation.

4. Microbiome-modulated immunity and campylobacteriosis

Because of their proximity, microbiome and gut immune system are actively interact with each other against the foreign substances and pathogens [11]. Gut microorganisms form a microbial community co-existed with the gut-associated lymphoid tissue [111], which is the largest immune organ in our body. Under normal circumstances, the intestinal epithelium and resident flora are separated by mucus layer, which not only provides static shielding, but also limits normal microbiomes' immunogenicity by imprinting dendritic cells [112, 113] that have ability to distinguish antigens present by normal microbiota and invaded pathogens [114]. Thus, the normal flora can live along with the host without causing damage, or getting removed by the host immunity [115]. The elimination of microbiomes results in a deficiency function of immunity, as a fact, antibiotics treated mice can be used as a model for the study of pathogen colonization [116]. Infectious pathogens often break gut microenvironment's equilibrium to generate ill effects, which may cause the gastrointestinal illnesses like campylobacteriosis. The normal flora have the capacity to induce lymphoid tissue's immune response to protect host from pathogens infection [117].

As one of the enteritis, campylobacteriosis has a common feature of leading extensive intestinal inflammation driven by Th1 and Th17 lymphocytes and TLR4 when homeostatic is perturbed [118], sharing typical pathology at cellular levels, such as neutrophils infiltration, leukocytes existence in fecal, and crypt abscesses. However, the pathogenesis of campylobacteriosis is not well studied. *C. jejuni* is commensalism with chickens [111], but causes diseases in humans [119]. Increasing data show that microbiomes play a pivotal role in modulating host immunity against campylobacteriosis and other enteritis. Better understanding the complex interaction between gut microbiome, pathogen *C. jejuni*, and host immune response is crucial for discovering new therapies to prevent and treat campylobacteriosis.

4.1 Immunity, microbiota, and *C. jejuni* interaction

The gut homeostasis is dependent on the symbiotic relationship interacts between microbiota and immunity, with the occasional breaks by intestinal diseases such as IBD and campylobacteriosis (**Figure 3**). Signals derived from gut microbiota are essential for the development of the immune system. Germ-free mice display impaired immunity maturation such as defective Peyer's patches (PPs), plasma cells, intraepithelial lymphocytes (IELs), antimicrobial peptide, IgA secretion, epithelial barrier function, and CD4+ T cell maturation [120–122]. Comparably, manipulating microbiota by antibiotic treatment or microbiota reconstitution (fecal microbiota transplantation, FMT) shows the essential role of the microbiota in immune homeostasis. FMT reduces dextran sulfate sodium (DSS)-induced mouse colitis with reduced $CD4^+$ T, $CD8^+$ T cells expressing, CD107a, MHC II-expressing, professional antigen present cells (APCs) expressing, while innate lymphocytes ILC2 and ILC3 are increased [123]. Human FMT to mice fails to resist *Salmonella* infection and restore the low levels of $CD4^+$ and $CD8^+$ T cells, proliferating T cells, dendritic cells, and antimicrobial peptide expression compared to mouse FMT

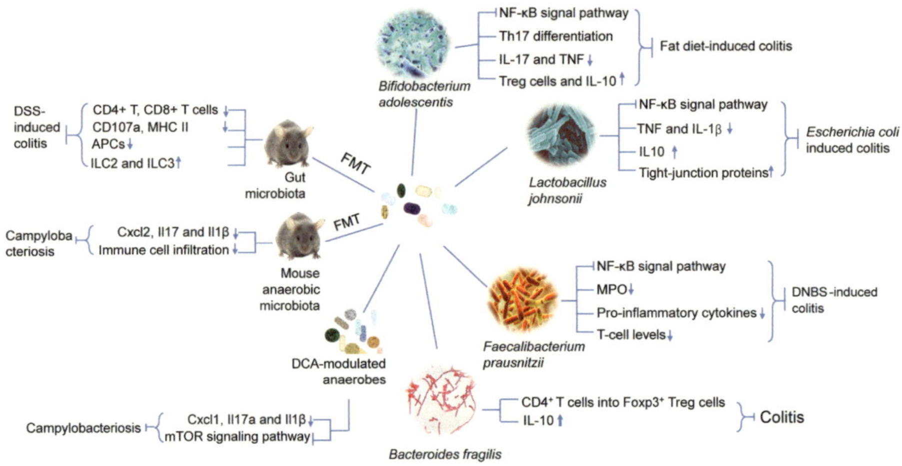

Figure 3.
Schematic illustration of the interaction of microbiota and immunity in campylobacteriosis.

[124], suggesting that gut immune maturation is dependent on colonization with a host-specific microbiota. Fecal Microbiota Transplantation (FMT) is successfully conducted on *Clostridium difficile* infection (CDI) patients by providing them with microbiomes from healthy donors to rebuilt the gut immunity [125] by inhibiting the activity of T cells and Th1 differentiation, preventing leukocyte adhesion, and production of inflammatory factors [126]. Consistently, ILC2-secreted IL-33 is essential for eosinophilia and tissue repair and survival, and its secretion is dependent on microbiota and can be rescued with FMT therapy to reduce *Clostridium difficile* infection (CDI) [127]. FMT of mouse anaerobic microbiota to germ free *Il10*$^{-/-}$ mice prevents *C. jejuni*-induced intestinal inflammation with reduced inflammatory genes of *Cxcl2*, *Il17* and *Il1β* as well as massive immune cell infiltration into gut lamina propria [85]. DCA-modulated anaerobes could attenuate chicken transmission exacerbated campylobacteriosis in mice by reduction of inflammatory genes, *Il17a*, *Il1β*, and *Cxcl1* expression in cellular level and inhibiting mTOR signaling pathway [128].

Beside microbiota transplantation, individual or groups of probiotics have been studied to reduce enteric pathogens, such as *Lactobacillus helveticus* [129], *Lactobacillus rhamnosus* LGG [130], *Lactobacillus gasseri* SBT2055 [131], *Lactobacillus* strains N8, N9, ZL4 and ZL5 [132], *Bifidobacterium longum* infantis [133]. *Lactobacillus* enhances macrophage elimination of *C. jejuni in vitro* and increases the expression of *Il1β*, *Il12p40*, *Il10*, and *Cxcl2* and the co-stimulatory molecules CD40, CD80, and CD86 [134]. Oral gavage of *Lactobacillus johnsonii* CJLJ103 inhibits LPS-induced NF-κB activation and *Tnfα* and *Il1β* expression, while expression of IL-10 and tight-junction proteins was increased [135]. *Lactobacillus plantarum* LC27 and *Bifidobacterium longum LC67* inhibits LPS, or 2,4,6-trinitrobenzesulfonic acid (TNBS)-induced colitis by suppress of NF-κB activation, CXCL4 expression and restored Th17/Treg balance [136]. Studies have shown that filamentous bacilli closely adhered to intestinal epithelium can induce Th17 reaction and increase the number of the anti-inflammatory Treg cells in the colon, and single colonization of *Bacillus fragilis* possesses immunomodulatory molecule-polysaccharide A (PSA), facilitates IL-10 producing through the conversion of CD4$^+$ T cells into Foxp3$^+$ Treg cells [137], and plays an important role in preventing and treatment of colitis in animals [138]. *Faecalibacterium prausnitzii* was reduced in patients with Crohn's disease [139]. *F. prausnitzii* supplementation prevents dinitrobenzene sulfonic acid (DNBS)-induced mouse colitis with inactivation of NF-κB signal pathway, down-regulation of MPO, pro-inflammatory cytokines, and T-cell levels [140]. With the advanced research on microbiota and immunity, it is expected that individual bacteria or groups of bacteria will be used to control *C. jeuni* infection in the near future.

4.2 Immunity, microbial metabolite, and *C. jejuni* interaction

In addition to the direct talk between microbiota and gut immunity, microbiota metabolites influence intestinal immune homeostasis, which is dependent on the balance of pro- and anti-inflammatory response (**Figure 4**). As discussed in section 2.2, Treg is the key ant-inflammatory T cell with its signature cytokine IL-10. IBD patients show reduced SCFAs in stool compared to healthy people, a consistent observation with reduced butyrate-producing bacterial taxa [141]. SCFAs, such as butyrate, acetate, and propionate, are microbial fermentation products of polysaccharides [142]. SCFAs are the energy source for colonocytes that lining the gastrointestinal tract [143], which have antiproliferative and anti-inflammatory features [144, 145]. SCFAs promotes the differentiation of Treg cells and their anti-inflammatory IL-10 secretion [146]. Butyrate or mixtures of SCFAs in enemas show clinical and histological improvement in active UC patients and diversion colitis [147, 148]. At the molecular level, butyrate in enemas decrease NF-κB activation in macrophages from distal colon tissue of UC patients [149], and reduce LPS-induced cytokine expression, NF-κB activation in lamina propria, and the number of peripheral blood monocytes in CD patients [150]. *C. jejuni* expresses the highest levels of the SCFA-related genes (*ggt*, *peb1c*, and *Cjj0683*) and colonizes efficiently in SCFAs-rich chicken ceca compared to other intestinal segments [151]. It remains elusive what is the role of SCFAs on *C. jejuni*-induced campylobacteriosis.

Besides microbiota metabolites regulation immune cells, they also modulate immune signaling pathways. Caffeic acid (CaA) is a hydrolyzed metabolite of chlorogenic acid by gut microbial esterase. CaA reduces DSS-induced in C57BL/6 mice colitis through blocking NF-κB signaling pathway, suppressing the secretion of IL-6, TNFα, and IFNγ, and inhibiting the infiltration of $CD3^+$ T cells, $CD177^+$ neutrophils and $F4/80^+$ macrophages [152]. L-arabinose, the digestion production of fiber, inhibits DSS-induced colitis by downregulating p38−/p65-dependent inflammation activation [153]. β-glucan is a polysaccharide naturally appeared in the cell walls of cereals, bacteria, and fungi. β-glucan reduces DSS-induced IBD by downregulating pro-inflammatory cytokines (TNFα, IL-6 and IL-8) and

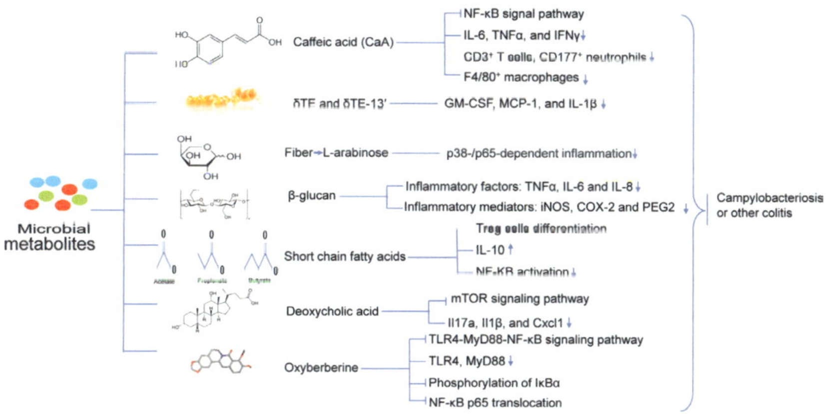

Figure 4.
Schematic illustration of the role of microbiota metabolites and immunity in campylobacteriosis.

inflammatory mediators (iNOS, COX-2 and PEG2) [154]. Oxyberberine, a gut microbiota metabolite of berberine, shown anti-colitis effect through the inhibition of TLR4-MyD88-NF-κB signaling pathway with reducing phosphorylation of IκBα and translocation of NF-κB p65 from cytoplasm to nucleus [155]. Notably, microbiota metabolic product DCA reduces *C. jejuni*-induced intestinal inflammation in $Il10^{-/-}$ mice with reduced inflammatory genes of *Cxcl2*, *Il17* and *Il1β* as well as massive immune cell infiltration into gut lamina propria [85]. Increasing microbiota metabolites will be discovered to attenuate *C. jejuni*-induced campylobacteriosis.

Furthermore, microbiota mediated metabolites are the important nutrients for host growth and immunity. Germ-free mice are usually more susceptible to infection diseases and show deficient to Vitamin K and B6 [156, 157]. Gut microbiota-synthesized Vitamins B12 and folate are vital for red blood cells synthesis, and red blood cells are crucial for supplying oxygen to immune cells and participating in the defensive process against pathogens [158]. Vitamin E delta-tocotrienol and its metabolite 13′-carboxychromanol inhibit tumor-associated colitis by reduction of pro-inflammatory cytokines GM-CSF, MCP-1, and IL-1β, respectively [159].

5. Conclusion

Given the fast research advancement on mucosal immunology, microbiota, and metabolomics recently in gastroenterology field, it is better than ever to investigate the mechanism of immunity-microbiota interaction and to use the knowledge to prevent and treat campylobacteriosis. The gut adaptive and innate system is the key for the permission or resistance to enteric pathogens and their induction of intestinal inflammation. Microbiota and its metabolic products or metabolites are essential for preventing gut pathogen invasion and the enteritis. Together, the development and function of the intestinal immunity is modulated by intestinal microbiota and its metabolic activities and products. Indeed, microbiota reconstitution by FMT is able to prevent or treat a number of intestinal disorders such as human CDI and mouse campylobacteriosis. Consistently, supplementing microbial metabolite of secondary bile acid DCA prevents campylobacteriosis in mice. Based on the successful or failed examples of the microbiome intervention on intestinal diseases, it is reasonable to conclude that a better knowledge on disease etiology and microbiome status during health and the diseases are essential for specifically targeting the pathogenic driving factors to prevent and treat the enteritis. Additional research will open new avenues to elucidate the in-depth understanding of the role of immunity and microbiota and to develop therapeutic approaches to control enteritis such as campylobacteriosis.

Funding

This research was supported by grants of Arkansas Biosciences Institute, USDA National Institute of Food and Agriculture (NIFA) Hatch project 1012366, NIFA Hatch/Multi State project 1018699, NIFA project 2020-67016-31346, and NIFA SAS 2019-69012-29905 to X. Sun. Poultry Federation Scholarships to Y. Fu. The funders had no role in study design, data collection and analysis, decision to publish, or preparation of the manuscript.

Conflict of interest

The authors declare no conflict of interest.

Author details

Ying Fu[1], Tahrir Alenezi[1], Ayidh Almansour[1], Hong Wang[1], Zhenquan Jia[2] and Xiaolun Sun[1]*

1 CEMB and Department of Poultry Science, University of Arkansas, Fayetteville, AR, USA

2 Department of Biology, University of North Carolina at Greensboro, Greensboro, NC, USA

*Address all correspondence to: xiaoluns@uark.edu

References

[1] F. Reich and G. Klein, "Legal aspects and microbiological criteria for Campylobacter spp. in the food processing chain," in *Campylobacter*: Elsevier, 2017, pp. 131-142.

[2] J. Silva, D. Leite, M. Fernandes, C. Mena, P. A. Gibbs, and P. Teixeira, "Campylobacter spp. as a Foodborne Pathogen: A Review," Front Microbiol, vol. 2, p. 200, 2011, doi: 10.3389/fmicb.2011.00200.

[3] J. Dicksved, P. Ellström, L. Engstrand, and H. Rautelin, "Susceptibility to Campylobacter infection is associated with the species composition of the human fecal microbiota," *MBio,* vol. 5, no. 5, 2014.

[4] CDC. https://www.cdc.gov/campylobacter/faq.html#:~:text=In%20 2015%2C%20National%20 Antimicrobial%20Resistance,makes%20 milk%20safe%20to%20drink. (accessed.

[5] CDC. https://www.cdc.gov/campylobacter/index.html (accessed.

[6] R. L. Scharff, "Food Attribution and Economic Cost Estimates for Meat- and Poultry-Related Illnesses," *J Food Prot,* vol. 83, no. 6, pp. 959-967, Jun 1 2020, doi: 10.4315/JFP-19-548.

[7] W. O. Masanta *et al.*, "Modification of intestinal microbiota and its consequences for innate immune response in the pathogenesis of campylobacteriosis," *Clinical and Developmental Immunology,* vol. 2013, 2013.

[8] M. Z. Cader and A. Kaser, "Recent advances in inflammatory bowel disease: mucosal immune cells in intestinal inflammation," Gut, vol. 62, no. 11, pp. 1653-1664, Nov 2013, doi: 10.1136/gutjnl-2012-303955.

[9] E. Dekaboruah, M. V. Suryavanshi, D. Chettri, and A. K. Verma, "Human microbiome: an academic update on human body site specific surveillance and its possible role," Arch Microbiol, vol. 202, no. 8, pp. 2147-2167, Oct 2020, doi: 10.1007/s00203-020-01931-x.

[10] I. Rowland *et al.*, "Gut microbiota functions: metabolism of nutrients and other food components," Eur J Nutr, vol. 57, no. 1, pp. 1-24, Feb 2018, doi: 10.1007/s00394-017-1445-8.

[11] M. G. Rooks and W. S. Garrett, "Gut microbiota, metabolites and host immunity," *Nat Rev Immunol,* vol. 16, no. 6, pp. 341-52, May 27 2016, doi: 10.1038/nri.2016.42.

[12] D. D. Chaplin, "Overview of the immune response," J Allergy Clin Immunol, vol. 125, no. 2 Suppl 2, pp. S3-23, Feb 2010, doi: 10.1016/j.jaci.2009.12.980.

[13] S. Akira, S. Uematsu, and O. Takeuchi, "Pathogen recognition and innate immunity," *Cell,* vol. 124, no. 4, pp. 783-801, Feb 24 2006, doi: 10.1016/j.cell.2006.02.015.

[14] X. Sun, D. Threadgill, and C. Jobin, "Campylobacter jejuni induces colitis through activation of mammalian target of rapamycin signaling," *Gastroenterology,* vol. 142, no. 1, pp. 86-95 e5, Jan 2012, doi: 10.1053/j.gastro.2011.09.042.

[15] M. F. Neurath, "Targeting immune cell circuits and trafficking in inflammatory bowel disease," Nat Immunol, vol. 20, no. 8, pp. 970-979, Aug 2019, doi: 10.1038/s41590-019-0415-0.

[16] R. Okamoto and M. Watanabe, "Role of epithelial cells in the pathogenesis and treatment of inflammatory bowel disease," J Gastroenterol, vol. 51,

no. 1, pp. 11-21, Jan 2016, doi: 10.1007/s00535-015-1098-4.

[17] M. Yamamoto, M. I. Pinto-Sanchez, P. Bercik, and P. Britz-McKibbin, "Metabolomics reveals elevated urinary excretion of collagen degradation and epithelial cell turnover products in irritable bowel syndrome patients," *Metabolomics,* vol. 15, no. 6, p. 82, May 20 2019, doi: 10.1007/s11306-019-1543-0.

[18] N. Bhutiani *et al.*, "Enhanced gut barrier integrity sensitizes colon cancer to immune therapy," *Oncoimmunology,* vol. 7, no. 11, p. e1498438, 2018, doi: 10.1080/2162402X.2018.1498438.

[19] G. Hecht, C. Pothoulakis, J. T. LaMont, and J. L. Madara, "Clostridium difficile toxin A perturbs cytoskeletal structure and tight junction permeability of cultured human intestinal epithelial monolayers," J Clin Invest, vol. 82, no. 5, pp. 1516-1524, Nov 1988, doi: 10.1172/JCI113760.

[20] E. Lippert *et al.*, "Gnotobiotic IL-10; NF-kappaB mice develop rapid and severe colitis following Campylobacter jejuni infection," *PLoS One,* vol. 4, no. 10, p. e7413, Oct 20 2009, doi: 10.1371/journal.pone.0007413.

[21] D. Hirayama, T. Iida, and H. Nakase, "The Phagocytic Function of Macrophage-Enforcing Innate Immunity and Tissue Homeostasis," *Int J Mol Sci,* vol. 19, no. 1, Dec 29 2017, doi: 10.3390/ijms19010092.

[22] T. M. Wassenaar, M. Engelskirchen, S. Park, and A. Lastovica, "Differential uptake and killing potential of Campylobacter jejuni by human peripheral monocytes/macrophages," Med Microbiol Immunol, vol. 186, no. 2-3, pp. 139-144, Oct 1997, doi: 10.1007/s004300050056.

[23] S. von Vietinghoff and K. Ley, "Homeostatic regulation of blood neutrophil counts," *J Immunol,* vol. 181, no. 8, pp. 5183-8, Oct 15 2008, doi: 10.4049/jimmunol.181.8.5183.

[24] X. Sun, B. Liu, R. B. Sartor, and C. Jobin, "Phosphatidylinositol 3-kinase-gamma signaling promotes Campylobacter jejuni-induced colitis through neutrophil recruitment in mice," *J Immunol,* vol. 190, no. 1, pp. 357-65, Jan 1 2013, doi: 10.4049/jimmunol.1201825.

[25] Y. Tang *et al.*, "Innate lymphoid cell composition shift upon Campylobacter jejuni induced colitis, a process inhibited by targeting mTOR signaling in Il10 mice," The Journal of Immunology, vol. 200, no. 1 Supplement, pp. 114.4-114.4, 2018.

[26] A. Hameed, "Human Immunity Against Campylobacter Infection," *Immune Netw,* vol. 19, no. 6, p. e38, Dec 2019, doi: 10.4110/in.2019.19.e38.

[27] J. M. Wells, O. Rossi, M. Meijerink, and P. van Baarlen, "Epithelial crosstalk at the microbiota-mucosal interface," *Proc Natl Acad Sci U S A,* vol. 108 Suppl 1, pp. 4607-14, Mar 15 2011, doi: 10.1073/pnas.1000092107.

[28] T. H. Mogensen, "Pathogen recognition and inflammatory signaling in innate immune defenses," *Clin Microbiol Rev,* vol. 22, no. 2, pp. 240-73, Table of Contents, Apr 2009, doi: 10.1128/CMR.00046-08.

[29] J. E. Shin *et al.*, "Lipo-oligosaccharides of Campylobacter jejuni serotype O:10. Structures of core oligosaccharide regions from a bacterial isolate from a patient with the Miller-Fisher syndrome and from the serotype reference strain," Carbohydr Res, vol. 305, no. 2, pp. 223-232, Dec 1997, doi: 10.1016/s0008-6215(97)00259-0.

[30] X. Sun, B. Allard, and C. Jobin, "MyD88/NF-κB Dependent Campylobacter Jejuni-Induced IL-12p40

Gene Expression Is Negatively Regulated By the AKT/GSK-3β Signaling Pathway in Murine Bone Marrow-Derived Dendritic Cells," *Gastroenterology,* vol. 136, no. 5, pp. Supplment 1, A41, 2009, doi: https://doi.org/10.1016/S0016-5085(09)60187-6.

[31] X. Sun and C. Jobin, "Nucleotide-binding oligomerization domain-containing protein 2 controls host response to Campylobacter jejuni in Il10−/− mice," *J Infect Dis,* vol. 210, no. 7, pp. 1145-54, Oct 1 2014, doi: 10.1093/infdis/jiu148.

[32] J. Deguine and G. M. Barton, "MyD88: a central player in innate immune signaling," *F1000Prime Rep,* vol. 6, p. 97, 2014, doi: 10.12703/P6-97.

[33] T. Kawasaki and T. Kawai, "Toll-like receptor signaling pathways," Front Immunol, vol. 5, p. 461, 2014, doi: 10.3389/fimmu.2014.00461.

[34] J. S. Roh and D. H. Sohn, "Damage-Associated Molecular Patterns in Inflammatory Diseases," *Immune Netw,* vol. 18, no. 4, p. e27, Aug 2018, doi: 10.4110/in.2018.18.e27.

[35] P. Broz and V. M. Dixit, "Inflammasomes: mechanism of assembly, regulation and signalling," Nat Rev Immunol, vol. 16, no. 7, pp. 407-420, Jul 2016, doi: 10.1038/nri.2016.58.

[36] M. D. Cooper, R. D. Peterson, and R. A. Good, "Delineation of the Thymic and Bursal Lymphoid Systems in the Chicken," Nature, vol. 205, pp. 143-146, Jan 9 1965, doi: 10.1038/205143a0.

[37] J. P. Cannon, R. N. Haire, J. P. Rast, and G. W. Litman, "The phylogenetic origins of the antigen-binding receptors and somatic diversification mechanisms," Immunol Rev, vol. 200, pp. 12-22, Aug 2004, doi: 10.1111/j.0105-2896.2004.00166.x.

[38] R. N. Germain, "T-cell development and the CD4-CD8 lineage decision," Nat Rev Immunol, vol. 2, no. 5, pp. 309-322, May 2002, doi: 10.1038/nri798.

[39] V. Brucklacher-Waldert, E. J. Carr, M. A. Linterman, and M. Veldhoen, "Cellular Plasticity of CD4+ T Cells in the Intestine," Front Immunol, vol. 5, p. 488, 2014, doi: 10.3389/fimmu.2014.00488.

[40] L. S. Kreisman and B. A. Cobb, "Glycoantigens induce human peripheral Tr1 cell differentiation with gut-homing specialization," *The Journal of biological chemistry,* vol. 286, no. 11, pp. 8810-8, Mar 18 2011, doi: 10.1074/jbc.M110.206011.

[41] B. Johansson-Lindbom *et al.*, "Functional specialization of gut CD103+ dendritic cells in the regulation of tissue-selective T cell homing," *The Journal of experimental medicine,* vol. 202, no. 8, pp. 1063-73, Oct 17 2005, doi: 10.1084/jem.20051100.

[42] C. M. Sun *et al.*, "Small intestine lamina propria dendritic cells promote de novo generation of Foxp3 T reg cells via retinoic acid," *The Journal of experimental medicine,* vol. 204, no. 8, pp. 1775-85, Aug 6 2007, doi: 10.1084/jem.20070602.

[43] J. J. Worthington, B. I. Czajkowska, A. C. Melton, and M. A. Travis, "Intestinal dendritic cells specialize to activate transforming growth factor-beta and induce Foxp3+ regulatory T cells via integrin alphavbeta8," Gastroenterology, vol. 141, no. 5, pp. 1802-1812, Nov 2011, doi: 10.1053/j.gastro.2011.06.057.

[44] L. Hu, M. D. Bray, M. Osorio, and D. J. Kopecko, "Campylobacter jejuni induces maturation and cytokine production in human dendritic cells," Infect Immun, vol. 74, no. 5, pp. 2697-2705, May 2006, doi: 10.1128/IAI.74.5.2697-2705.2006.

[45] M. Bax *et al.*, "Campylobacter jejuni lipooligosaccharides modulate dendritic

cell-mediated T cell polarization in a sialic acid linkage-dependent manner," Infect Immun, vol. 79, no. 7, pp. 2681-2689, Jul 2011, doi: 10.1128/IAI.00009-11.

[46] S. Li *et al.*, "Circulating Th17, Th22, and Th1 cells are elevated in the Guillain-Barre syndrome and downregulated by IVIg treatments," Mediators Inflamm, vol. 2014, p. 740947, 2014, doi: 10.1155/2014/740947.

[47] M. A. Strid, J. Engberg, L. B. Larsen, K. Begtrup, K. Molbak, and K. A. Krogfelt, "Antibody responses to Campylobacter infections determined by an enzyme-linked immunosorbent assay: 2-year follow-up study of 210 patients," Clin Diagn Lab Immunol, vol. 8, no. 2, pp. 314-319, Mar 2001, doi: 10.1128/CDLI.8.2.314-319.2001.

[48] S. A. Cawthraw, R. A. Feldman, A. R. Sayers, and D. G. Newell, "Long-term antibody responses following human infection with Campylobacter jejuni," Clin Exp Immunol, vol. 130, no. 1, pp. 101-106, Oct 2002, doi: 10.1046/j.1365-2249.2002.01966.x.

[49] S. A. Cawthraw, L. Lind, B. Kaijser, and D. G. Newell, "Antibodies, directed towards Campylobacter jejuni antigens, in sera from poultry abattoir workers," Clin Exp Immunol, vol. 122, no. 1, pp. 55-60, Oct 2000, doi: 10.1046/j.1365-2249.2000.01349.x.

[50] R. S. Houliston *et al.*, "Lipooligosaccharide of Campylobacter jejuni: similarity with multiple types of mammalian glycans beyond gangliosides," *J Biol Chem*, vol. 286, no. 14, pp. 12361-70, Apr 8 2011, doi: 10.1074/jbc.M110.181750.

[51] P. J. Turnbaugh, R. E. Ley, M. Hamady, C. M. Fraser-Liggett, R. Knight, and J. I. Gordon, "The human microbiome project," Nature, vol. 449, no. 7164, pp. 804-810, 2007.

[52] X. Sun and Z. Jia, "Microbiome modulates intestinal homeostasis against inflammatory diseases," Veterinary immunology and immunopathology, vol. 205, pp. 97-105, Nov 2018, doi: 10.1016/j.vetimm.2018.10.014.

[53] M. Blaut and T. Clavel, "Metabolic diversity of the intestinal microbiota: implications for health and disease," The Journal of nutrition, vol. 137, no. 3, pp. 751S–755S, 2007.

[54] E. Holmes, J. V. Li, T. Athanasiou, H. Ashrafian, and J. K. Nicholson, "Understanding the role of gut microbiome–host metabolic signal disruption in health and disease," Trends in microbiology, vol. 19, no. 7, pp. 349-359, 2011.

[55] T. D. Lawley and A. W. Walker, "Intestinal colonization resistance," Immunology, vol. 138, no. 1, pp. 1-11, 2013.

[56] P. Hugon, J. C. Dufour, P. Colson, P. E. Fournier, K. Sallah, and D. Raoult, "A comprehensive repertoire of prokaryotic species identified in human beings," Lancet Infect Dis, vol. 15, no. 10, pp. 1211-1219, Oct 2015, doi: 10.1016/S1473-3099(15)00293-5.

[57] G. E. Diehl *et al.*, "Microbiota restricts trafficking of bacteria to mesenteric lymph nodes by CX 3 CR1 hi cells," Nature, vol. 494, no. 7435, pp. 116-120, 2013.

[58] L. Lu and W. A. Walker, "Pathologic and physiologic interactions of bacteria with the gastrointestinal epithelium," The American journal of clinical nutrition, vol. 73, no. 6, pp. 1124S–1130S, 2001.

[59] W. R. Wikoff *et al.*, "Metabolomics analysis reveals large effects of gut microflora on mammalian blood metabolites," Proceedings of the national academy of sciences, vol. 106, no. 10, pp. 3698-3703, 2009.

[60] D. Dodd *et al.*, "A gut bacterial pathway metabolizes aromatic amino acids into nine circulating metabolites," Nature, vol. 551, no. 7682, pp. 648-652, 2017.

[61] F.-P. J. Martin *et al.*, "Panorganismal gut microbiome– host metabolic crosstalk," Journal of proteome research, vol. 8, no. 4, pp. 2090-2105, 2009.

[62] S. P. Claus *et al.*, "Systemic multicompartmental effects of the gut microbiome on mouse metabolic phenotypes," Molecular systems biology, vol. 4, no. 1, p. 219, 2008.

[63] J. R. Lupton, "Microbial degradation products influence colon cancer risk: the butyrate controversy," The Journal of nutrition, vol. 134, no. 2, pp. 479-482, 2004.

[64] D. R. Donohoe, L. B. Collins, A. Wali, R. Bigler, W. Sun, and S. J. Bultman, "The Warburg effect dictates the mechanism of butyrate-mediated histone acetylation and cell proliferation," Molecular cell, vol. 48, no. 4, pp. 612-626, 2012.

[65] A. Wahlström, S. I. Sayin, H.-U. Marschall, and F. Bäckhed, "Intestinal crosstalk between bile acids and microbiota and its impact on host metabolism," Cell metabolism, vol. 24, no. 1, pp. 41-50, 2016.

[66] M. S. Desai *et al.*, "A dietary fiber-deprived gut microbiota degrades the colonic mucus barrier and enhances pathogen susceptibility," *Cell*, vol. 167, no. 5, pp. 1339-1353. e21, 2016.

[67] A. Koh, F. De Vadder, P. Kovatcheva-Datchary, and F. Bäckhed, "From dietary fiber to host physiology: short-chain fatty acids as key bacterial metabolites," Cell, vol. 165, no. 6, pp. 1332-1345, 2016.

[68] S. Macfarlane and G. T. Macfarlane, "Regulation of short-chain fatty acid production," Proceedings of the Nutrition Society, vol. 62, no. 1, pp. 67-72, 2003.

[69] K. Makki, E. C. Deehan, J. Walter, and F. Bäckhed, "The impact of dietary fiber on gut microbiota in host health and disease," Cell host & microbe, vol. 23, no. 6, pp. 705-715, 2018.

[70] J. M. Ridlon, D.-J. Kang, and P. B. Hylemon, "Bile salt biotransformations by human intestinal bacteria," Journal of lipid research, vol. 47, no. 2, pp. 241-259, 2006.

[71] A. F. Hofmann, "The continuing importance of bile acids in liver and intestinal disease," Archives of internal medicine, vol. 159, no. 22, pp. 2647-2658, 1999.

[72] M. C. Rea *et al.*, "Thuricin CD, a posttranslationally modified bacteriocin with a narrow spectrum of activity against Clostridium difficile," Proceedings of the National Academy of Sciences, vol. 107, no. 20, pp. 9352-9357, 2010.

[73] S. Rebuffat, "Bacteriocins from Gram-negative bacteria: a classification?," in *Prokaryotic antimicrobial peptides*: Springer, 2011, pp. 55-72.

[74] M. C. Rea, R. P. Ross, P. D. Cotter, and C. Hill, "Classification of bacteriocins from Gram-positive bacteria," in *Prokaryotic antimicrobial peptides*: Springer, 2011, pp. 29-53.

[75] W. M. Parks, A. R. Bottrill, O. A. Pierrat, M. C. Durrant, and A. Maxwell, "The action of the bacterial toxin, microcin B17, on DNA gyrase," Biochimie, vol. 89, no. 4, pp. 500-507, 2007.

[76] D. Destoumieux-Garzón, J. Peduzzi, X. Thomas, C. Djediat, and S. Rebuffat, "Parasitism of iron-siderophore receptors of Escherichia coli by the

siderophore-peptide microcin E492m and its unmodified counterpart," Biometals, vol. 19, no. 2, pp. 181-191, 2006.

[77] J. Mukhopadhyay, E. Sineva, J. Knight, R. M. Levy, and R. H. Ebright, "Antibacterial peptide microcin J25 inhibits transcription by binding within and obstructing the RNA polymerase secondary channel," Molecular cell, vol. 14, no. 6, pp. 739-751, 2004.

[78] P. D. Cotter, R. P. Ross, and C. Hill, "Bacteriocins—a viable alternative to antibiotics?," Nature Reviews Microbiology, vol. 11, no. 2, pp. 95-105, 2013.

[79] B. Alrubaye *et al.*, "Microbial metabolite deoxycholic acid shapes microbiota against Campylobacter jejuni chicken colonization," *PloS one,* vol. 14, no. 7, p. e0214705, 2019.

[80] C. Pielsticker, G. Glunder, and S. Rautenschlein, "Colonization properties of Campylobacter jejuni in chickens," Eur J Microbiol Immunol (Bp), vol. 2, no. 1, pp. 61-65, Mar 2012, doi: 10.1556/EuJMI.2.2012.1.9.

[81] C. Chang and J. F. Miller, "Campylobacter jejuni colonization of mice with limited enteric flora," Infect Immun, vol. 74, no. 9, pp. 5261-5271, Sep 2006, doi: 10.1128/IAI.01094-05.

[82] M. J. Blaser and W. L. Wang, "Campylobacter infections in human beings," *J Pediatr,* vol. 96, no. 2, p. 343, Feb 1980, doi: 10.1016/s0022-3476(80)80844-4.

[83] I. Patuzzi *et al.*, "The Interplay between Campylobacter and the Caecal Microbial Community of Commercial Broiler Chickens over Time," *Microorganisms,* vol. 9, no. 2, Jan 22 2021, doi: 10.3390/microorganisms9020221.

[84] N. O. Kaakoush, N. Sodhi, J. W. Chenu, J. M. Cox, S. M. Riordan, and H. M. Mitchell, "The interplay between Campylobacter and Helicobacter species and other gastrointestinal microbiota of commercial broiler chickens," Gut Pathogens, vol. 6, no. 1, p. 18, 2014.

[85] X. Sun *et al.*, "Microbiota-Derived Metabolic Factors Reduce Campylobacteriosis in Mice," *Gastroenterology,* vol. 154, no. 6, pp. 1751-1763 e2, May 2018, doi: 10.1053/j.gastro.2018.01.042.

[86] J. L. O'Loughlin *et al.*, "The Intestinal Microbiota Influences Campylobacter jejuni Colonization and Extraintestinal Dissemination in Mice," Appl Environ Microbiol, vol. 81, no. 14, pp. 4642-4650, Jul 2015, doi: 10.1128/AEM.00281-15.

[87] L.-M. Haag *et al.*, "Intestinal microbiota shifts towards elevated commensal Escherichia coli loads abrogate colonization resistance against Campylobacter jejuni in mice," PloS one, vol. 7, no. 5, p. e35988, 2012.

[88] A. Metwaly *et al.*, "Integrated microbiota and metabolite profiles link Crohn's disease to sulfur metabolism," *Nat Commun,* vol. 11, no. 1, p. 4322, Aug 28 2020, doi: 10.1038/s41467-020-17956-1.

[89] D. Knights *et al.*, "Complex host genetics influence the microbiome in inflammatory bowel disease," *Genome Med,* vol. 6, no. 12, p. 107, 2014, doi: 10.1186/s13073-014-0107-1.

[90] X. C. Morgan *et al.*, "Dysfunction of the intestinal microbiome in inflammatory bowel disease and treatment," *Genome Biol,* vol. 13, no. 9, p. R79, Apr 16 2012, doi: 10.1186/gb-2012-13-9-r79.

[91] J. Wang, T. Lang, J. Shen, J. Dai, L. Tian, and X. Wang, "Core Gut Bacteria Analysis of Healthy Mice," Front Microbiol, vol. 10, p. 887, 2019, doi: 10.3389/fmicb.2019.00887.

[92] A. Cean *et al.*, "Effect of human isolated probiotic bacteria on preventing Campylobacter jejuni colonization of poultry," Foodborne pathogens and disease, vol. 12, no. 2, pp. 122-130, 2015.

[93] L. Baffoni, F. Gaggìa, D. Di Gioia, C. Santini, L. Mogna, and B. Biavati, "A Bifidobacterium-based synbiotic product to reduce the transmission of C. jejuni along the poultry food chain," International journal of food microbiology, vol. 157, no. 2, pp. 156-161, 2012.

[94] M. Ganan, A. J. Martinez-Rodriguez, A. V. Carrascosa, S. Vesterlund, S. Salminen, and R. Satokari, "Interaction of Campylobacter spp. and human probiotics in chicken intestinal mucus," Zoonoses and public health, vol. 60, no. 2, pp. 141-148, 2013.

[95] R. Tareb, M. Bernardeau, M. Gueguen, and J.-P. Vernoux, "In vitro characterization of aggregation and adhesion properties of viable and heat-killed forms of two probiotic Lactobacillus strains and interaction with foodborne zoonotic bacteria, especially Campylobacter jejuni," Journal of Medical Microbiology, vol. 62, no. 4, pp. 637-649, 2013.

[96] S. Messaoudi *et al.*, "In vitro evaluation of the probiotic potential of Lactobacillus salivarius SMXD51," Anaerobe, vol. 18, no. 6, pp. 584-589, 2012.

[97] C. Kampmann, J. Dicksved, L. Engstrand, and H. Rautelin, "Composition of human faecal microbiota in resistance to Campylobacter infection," *Clinical Microbiology and Infection,* vol. 22, no. 1, pp. 61. e1-61. e8, 2016.

[98] C. De Filippo *et al.*, "Impact of diet in shaping gut microbiota revealed by a comparative study in children from Europe and rural Africa," Proceedings of the National Academy of Sciences, vol. 107, no. 33, pp. 14691-14696, 2010.

[99] J. P. Jacobs *et al.*, "A Disease-Associated Microbial and Metabolomics State in Relatives of Pediatric Inflammatory Bowel Disease Patients," Cell Mol Gastroenterol Hepatol, vol. 2, no. 6, pp. 750-766, Nov 2016, doi: 10.1016/j.jcmgh.2016.06.004.

[100] E. A. Franzosa *et al.*, "Gut microbiome structure and metabolic activity in inflammatory bowel disease," Nat Microbiol, vol. 4, no. 2, pp. 293-305, Feb 2019, doi: 10.1038/s41564-018-0306-4.

[101] J. Lloyd-Price *et al.*, "Multi-omics of the gut microbial ecosystem in inflammatory bowel diseases," Nature, vol. 569, no. 7758, pp. 655-662, May 2019, doi: 10.1038/s41586-019-1237-9.

[102] H. Duboc *et al.*, "Connecting dysbiosis, bile-acid dysmetabolism and gut inflammation in inflammatory bowel diseases," Gut, vol. 62, no. 4, pp. 531-539, Apr 2013, doi: 10.1136/gutjnl-2012-302578.

[103] M. B. Miller and B. L. Bassler, "Quorum sensing in bacteria," Annu Rev Microbiol, vol. 55, pp. 165-199, 2001, doi: 10.1146/annurev.micro.55.1.165.

[104] J. E. Gonzalez and N. D. Keshavan, "Messing with bacterial quorum sensing," Microbiol Mol Biol Rev, vol. 70, no. 4, pp. 859-875, Dec 2006, doi: 10.1128/MMBR.00002-06.

[105] K. T. Elvers and S. F. Park, "Quorum sensing in Campylobacter jejuni: detection of a luxS encoded signalling molecule," Microbiology (Reading), vol. 148, no. Pt 5, pp. 1475-1481, May 2002, doi: 10.1099/00221287-148-5-1475.

[106] K. Bezek *et al.*, "Attenuation of Adhesion, Biofilm Formation and

Quorum Sensing of Campylobacter jejuni by Euodia ruticarpa," Phytother Res, vol. 30, no. 9, pp. 1527-1532, Sep 2016, doi: 10.1002/ptr.5658.

[107] G. Tram, C. J. Day, and V. Korolik, "Bridging the Gap: A Role for Campylobacter jejuni Biofilms," *Microorganisms,* vol. 8, no. 3, Mar 23 2020, doi: 10.3390/microorganisms8030452.

[108] B. Quinones, W. G. Miller, A. H. Bates, and R. E. Mandrell, "Autoinducer-2 production in Campylobacter jejuni contributes to chicken colonization," Appl Environ Microbiol, vol. 75, no. 1, pp. 281-285, Jan 2009, doi: 10.1128/AEM.01803-08.

[109] P. Plummer, J. Zhu, M. Akiba, D. Pei, and Q. Zhang, "Identification of a key amino acid of LuxS involved in AI-2 production in Campylobacter jejuni," *PLoS One,* vol. 6, no. 1, p. e15876, Jan 11 2011, doi: 10.1371/journal.pone.0015876.

[110] P. Plummer *et al.*, "Critical role of LuxS in the virulence of Campylobacter jejuni in a guinea pig model of abortion," Infect Immun, vol. 80, no. 2, pp. 585-593, Feb 2012, doi: 10.1128/IAI.05766-11.

[111] S. Humphrey *et al.*, "Campylobacter jejuni is not merely a commensal in commercial broiler chickens and affects bird welfare," *mBio,* vol. 5, no. 4, pp. e01364-14, Jul 1 2014, doi: 10.1128/mBio.01364-14.

[112] Y. Belkaid and S. Naik, "Compartmentalized and systemic control of tissue immunity by commensals," Nat Immunol, vol. 14, no. 7, pp. 646-653, Jul 2013, doi: 10.1038/ni.2604.

[113] M. Shan *et al.*, "Mucus enhances gut homeostasis and oral tolerance by delivering immunoregulatory signals," *Science,* vol. 342, no. 6157, pp. 447-53, Oct 25 2013, doi: 10.1126/science.1237910.

[114] A. J. Macpherson and T. Uhr, "Induction of protective IgA by intestinal dendritic cells carrying commensal bacteria," *Science,* vol. 303, no. 5664, pp. 1662-5, Mar 12 2004, doi: 10.1126/science.1091334.

[115] C. P. Davis, "Normal Flora," in *Medical Microbiology*, th and S. Baron Eds. Galveston (TX), 1996.

[116] S. Bereswill *et al.*, "Novel murine infection models provide deep insights into the "menage a trois" of Campylobacter jejuni, microbiota and host innate immunity," *PLoS One,* vol. 6, no. 6, p. e20953, 2011, doi: 10.1371/journal.pone.0020953.

[117] O. Pabst and A. M. Mowat, "Oral tolerance to food protein," Mucosal Immunol, vol. 5, no. 3, pp. 232-239, May 2012, doi: 10.1038/mi.2012.4.

[118] T. Imam, S. Park, M. H. Kaplan, and M. R. Olson, "Effector T Helper Cell Subsets in Inflammatory Bowel Diseases," Front Immunol, vol. 9, p. 1212, 2018, doi: 10.3389/fimmu.2018.01212.

[119] S. F. Altekruse, N. J. Stern, P. I. Fields, and D. L. Swerdlow, "Campylobacter jejuni--an emerging foodborne pathogen," Emerg Infect Dis, vol. 5, no. 1, pp. 28-35, Jan-Feb 1999, doi: 10.3201/eid0501.990104.

[120] H. Bauer, F. Paronetto, W. A. Burns, and A. Einheber, "The enhancing effect of the microbial flora on macrophage function and the immune response. A study in germfree mice," *J Exp Med,* vol. 123, no. 6, pp. 1013-24, Jun 1 1966, doi: 10.1084/jem.123.6.1013.

[121] E. M. Brown, D. J. Kenny, and R. J. Xavier, "Gut Microbiota

Regulation of T Cells During Inflammation and Autoimmunity," Annu Rev Immunol, vol. 37, pp. 599-624, Apr 26 2019, doi: 10.1146/annurev-immunol-042718-041841.

[122] J. L. Round and S. K. Mazmanian, "The gut microbiota shapes intestinal immune responses during health and disease," Nat Rev Immunol, vol. 9, no. 5, pp. 313-323, May 2009, doi: 10.1038/nri2515.

[123] C. Burrello *et al.*, "Fecal Microbiota Transplantation Controls Murine Chronic Intestinal Inflammation by Modulating Immune Cell Functions and Gut Microbiota Composition," *Cells,* vol. 8, no. 6, May 28 2019, doi: 10.3390/cells8060517.

[124] H. Chung *et al.*, "Gut immune maturation depends on colonization with a host-specific microbiota," *Cell,* vol. 149, no. 7, pp. 1578-93, Jun 22 2012, doi: 10.1016/j.cell.2012.04.037.

[125] C. R. Kelly *et al.*, "Fecal microbiota transplant for treatment of Clostridium difficile infection in immunocompromised patients," Am J Gastroenterol, vol. 109, no. 7, pp. 1065-1071, Jul 2014, doi: 10.1038/ajg.2014.133.

[126] Z. H. Shen *et al.*, "Relationship between intestinal microbiota and ulcerative colitis: Mechanisms and clinical application of probiotics and fecal microbiota transplantation," *World J Gastroenterol,* vol. 24, no. 1, pp. 5-14, Jan 7 2018, doi: 10.3748/wjg.v24.i1.5.

[127] A. L. Frisbee and W. A. Petri, Jr., "Considering the Immune System during Fecal Microbiota Transplantation for Clostridioides difficile Infection," Trends Mol Med, vol. 26, no. 5, pp. 496-507, May 2020, doi: 10.1016/j.molmed.2020.01.009.

[128] Y. Fu *et al.*, "Microbiota attenuates chicken transmission-exacerbated campylobacteriosis in Il10(−/−) mice," *Scientific reports,* vol. 10, no. 1, p. 20841, Nov 30 2020, doi: 10.1038/s41598-020-77789-2.

[129] E. Wine, M. G. Gareau, K. Johnson-Henry, and P. M. Sherman, "Strain-specific probiotic (Lactobacillus helveticus) inhibition of Campylobacter jejuni invasion of human intestinal epithelial cells," FEMS Microbiol Lett, vol. 300, no. 1, pp. 146-152, Nov 2009, doi: 10.1111/j.1574-6968.2009.01781.x.

[130] M. Sikic Pogacar, T. Langerholc, D. Micetic-Turk, S. S. Mozina, and A. Klancnik, "Effect of Lactobacillus spp. on adhesion, invasion, and translocation of Campylobacter jejuni in chicken and pig small-intestinal epithelial cell lines," *BMC Vet Res,* vol. 16, no. 1, p. 34, Feb 3 2020, doi: 10.1186/s12917-020-2238-5.

[131] K. Nishiyama *et al.*, "Lactobacillus gasseri SBT2055 reduces infection by and colonization of Campylobacter jejuni," *PLoS One,* vol. 9, no. 9, p. e108827, 2014, doi: 10.1371/journal.pone.0108827.

[132] G. Wang *et al.*, "Screening of adhesive lactobacilli with antagonistic activity against Campylobacter jejuni," *Food Control,* vol. 44, pp. 49-57, 2014/10/01/ 2014, doi: https://doi.org/10.1016/j.foodcont.2014.03.042.

[133] E. M. Quinn, H. Slattery, D. Walsh, L. Joshi, and R. M. Hickey, "Bifidobacterium longum subsp. infantis ATCC 15697 and Goat Milk Oligosaccharides Show Synergism In Vitro as Anti-Infectives against Campylobacter jejuni," *Foods,* vol. 9, no. 3, Mar 17 2020, doi: 10.3390/foods9030348.

[134] K. Taha-Abdelaziz *et al.*, "In vitro assessment of immunomodulatory and anti-Campylobacter activities of probiotic lactobacilli," *Sci Rep,* vol. 9, no. 1, p. 17903, Nov 29 2019, doi: 10.1038/s41598-019-54494-3.

[135] J. H. Lim SuMin, Jeong JinJu, Han MyungJoo, Kim DongHyun "Lactobacillus johnsonii CJLJ103 attenuates colitis and memory impairment in mice by inhibiting gut microbiota lipopolysaccharide production and NF-κB activation," Journal of Functional Foods, vol. 34, pp. 359-368, 2017.

[136] S. E. Jang, J. J. Jeong, J. K. Kim, M. J. Han, and D. H. Kim, "Simultaneous Amelioratation of Colitis and Liver Injury in Mice by Bifidobacterium longum LC67 and Lactobacillus plantarum LC27," *Sci Rep,* vol. 8, no. 1, p. 7500, May 14 2018, doi: 10.1038/s41598-018-25775-0.

[137] V. Gaboriau-Routhiau *et al.*, "The key role of segmented filamentous bacteria in the coordinated maturation of gut helper T cell responses," *Immunity,* vol. 31, no. 4, pp. 677-89, Oct 16 2009, doi: 10.1016/j.immuni.2009.08.020.

[138] J. L. Round and S. K. Mazmanian, "Inducible Foxp3+ regulatory T-cell development by a commensal bacterium of the intestinal microbiota," *Proc Natl Acad Sci U S A,* vol. 107, no. 27, pp. 12204-9, Jul 6 2010, doi: 10.1073/pnas.0909122107.

[139] E. Quevrain *et al.*, "Identification of an anti-inflammatory protein from Faecalibacterium prausnitzii, a commensal bacterium deficient in Crohn's disease," Gut, vol. 65, no. 3, pp. 415-425, Mar 2016, doi: 10.1136/gutjnl-2014-307649.

[140] R. Martin *et al.*, "The commensal bacterium Faecalibacterium prausnitzii is protective in DNBS-induced chronic moderate and severe colitis models," Inflamm Bowel Dis, vol. 20, no. 3, pp. 417-430, Mar 2014, doi: 10.1097/01.MIB.0000440815.76627.64.

[141] K. Machiels *et al.*, "A decrease of the butyrate-producing species Roseburia hominis and Faecalibacterium prausnitzii defines dysbiosis in patients with ulcerative colitis," Gut, vol. 63, no. 8, pp. 1275-1283, Aug 2014, doi: 10.1136/gutjnl-2013-304833.

[142] Y. P. Silva, A. Bernardi, and R. L. Frozza, "The Role of Short-Chain Fatty Acids From Gut Microbiota in Gut-Brain Communication," *Front Endocrinol (Lausanne),* vol. 11, p. 25, 2020, doi: 10.3389/fendo.2020.00025.

[143] D. Parada Venegas *et al.*, "Short Chain Fatty Acids (SCFAs)-Mediated Gut Epithelial and Immune Regulation and Its Relevance for Inflammatory Bowel Diseases," Front Immunol, vol. 10, p. 277, 2019, doi: 10.3389/fimmu.2019.00277.

[144] D. R. Donohoe *et al.*, "The microbiome and butyrate regulate energy metabolism and autophagy in the mammalian colon," *Cell Metab,* vol. 13, no. 5, pp. 517-26, May 4 2011, doi: 10.1016/j.cmet.2011.02.018.

[145] R. Correa-Oliveira, J. L. Fachi, A. Vieira, F. T. Sato, and M. A. Vinolo, "Regulation of immune cell function by short-chain fatty acids," *Clin Transl Immunology,* vol. 5, no. 4, p. e73, Apr 2016, doi: 10.1038/cti.2016.17.

[146] A. H. Keshteli, K. L. Madsen, and L. A. Dieleman, "Diet in the Pathogenesis and Management of Ulcerative Colitis; A Review of Randomized Controlled Dietary Interventions," *Nutrients,* vol. 11, no. 7, Jun 30 2019, doi: 10.3390/nu11071498.

[147] J. M. Harig, K. H. Soergel, R. A. Komorowski, and C. M. Wood, "Treatment of diversion colitis with short-chain-fatty acid irrigation," *N Engl J Med,* vol. 320, no. 1, pp. 23-8, Jan 5 1989, doi: 10.1056/NEJM198901053200105.

[148] W. Scheppach *et al.*, "Effect of butyrate enemas on the colonic

mucosa in distal ulcerative colitis," Gastroenterology, vol. 103, no. 1, pp. 51-56, Jul 1992, doi: 10.1016/0016-5085(92)91094-k.

[149] H. Luhrs *et al.*, "Butyrate inhibits NF-kappaB activation in lamina propria macrophages of patients with ulcerative colitis," Scand J Gastroenterol, vol. 37, no. 4, pp. 458-466, Apr 2002, doi: 10.1080/003655202317316105.

[150] J. P. Segain *et al.*, "Butyrate inhibits inflammatory responses through NFkappaB inhibition: implications for Crohn's disease," Gut, vol. 47, no. 3, pp. 397-403, Sep 2000, doi: 10.1136/gut.47.3.397.

[151] P. M. Luethy, S. Huynh, D. A. Ribardo, S. E. Winter, C. T. Parker, and D. R. Hendrixson, "Microbiota-Derived Short-Chain Fatty Acids Modulate Expression of Campylobacter jejuni Determinants Required for Commensalism and Virulence," *mBio,* vol. 8, no. 3, May 9 2017, doi: 10.1128/mBio.00407-17.

[152] Z. Zhang *et al.*, "Caffeic acid ameliorates colitis in association with increased Akkermansia population in the gut microbiota of mice," *Oncotarget,* vol. 7, no. 22, pp. 31790-9, May 31 2016, doi: 10.18632/oncotarget.9306.

[153] Y. Li *et al.*, "l-Arabinose Inhibits Colitis by Modulating Gut Microbiota in Mice," *J Agric Food Chem,* vol. 67, no. 48, pp. 13299-13306, Dec 4 2019, doi: 10.1021/acs.jafc.9b05829.

[154] H. F. Feifei Han, Ming Yao, Shasha Yang, Jianzhong Han, "Oral administration of yeast β-glucan ameliorates inflammation and intestinal barrier in dextran sodium sulfate-induced acute colitis," Journal of Functional Foods, vol. 35, pp. 115-126, 2017.

[155] C. Li *et al.*, "Oxyberberine, a novel gut microbiota-mediated metabolite of berberine, possesses superior anti-colitis effect: Impact on intestinal epithelial barrier, gut microbiota profile and TLR4-MyD88-NF-kappaB pathway," Pharmacol Res, vol. 152, p. 104603, Feb 2020, doi: 10.1016/j.phrs.2019.104603.

[156] B. E. Gustafsson, "Vitamin K deficiency in germfree rats," Ann N Y Acad Sci, vol. 78, pp. 166-174, May 8 1959, doi: 10.1111/j.1749-6632.1959.tb53101.x.

[157] Y. Sumi, M. Miyakawa, M. Kanzaki, and Y. Kotake, "Vitamin B-6 deficiency in germfree rats," J Nutr, vol. 107, no. 9, pp. 1707-1714, Sep 1977, doi: 10.1093/jn/107.9.1707.

[158] I. M. Bishlawy, "Red blood cells, hemoglobin and the immune system," Med Hypotheses, vol. 53, no. 4, pp. 345-346, Oct 1999, doi: 10.1054/mehy.1997.0778.

[159] C. Yang, Y. Zhao, S. Im, C. Nakatsu, Y. Jones-Hall, and Q. Jiang, "Vitamin E delta-tocotrienol and metabolite 13′-carboxychromanol inhibit colitis-associated colon tumorigenesis and modulate gut microbiota in mice," *J Nutr Biochem,* vol. 89, p. 108567, Jan 8 2021, doi: 10.1016/j.jnutbio.2020.108567.

Chapter 4

Isolation and Identification of *Campylobacter* species from Food and Food-Related Environment

Honsheng Huang and Manuel Mariano Garcia

Abstract

Campylobacter species are among the most common causes of bacterial gastroenteritis in humans worldwide. The genus *Campylobacter* consists of at least 39 validly published species with wide distribution in various hosts and environments, which are either pathogens for humans or animals, or not pathogenic as identified so far. Various methods have been used for detecting campylobacters including conventional culture methods, molecular (such as polymerase chain reaction), immunological methods and genome sequencing. Currently, isolation and subsequent identification of the target campylobacters are required by most of the regulatory bodies globally. The multiple *Campylobacter* species exhibit diverse physiological and metabolic characteristics and growth requirements, which can interfere with the sensitivity and specificity of culture-dependent methods. Furthermore, strains among each species may behavior differently in various culture media and under various culture conditions. Therefore, it is important to apply appropriate isolation and identification methods for different types of species and samples based on specific purposes. This chapter will review the development and the current status of culture-dependent methods for the isolation and detection of various *Campylobacter* species from food and food-related environments during the next generation sequencing era.

Keywords: campylobacter species, foodborne, physiological characteristics, isolation and identification

1. Introduction

1.1 Organisms and brief history

For the last three decades, *Campylobacter* species have been the focus of growing attention because of the increasing frequency with which they have been isolated from various sources including man, animals, food and water [1]. *Campylobacter* species are among the most common causes of bacterial gastroenteritis in human worldwide [2–4], which account annually for approximately 166 million foodborne illnesses around the world [5].

The name "*Campylobacter*" is an ancient Greek word meaning "curved rod". *Campylobacter* species are Gram-negative, spiral, rod-shaped, or curved bacteria with a single polar flagellum, bipolar flagella, or no flagellum depending on the species, non-spore-forming, and approximately 0.2 to 0.8 by 0.5 to 5 μm [6–8].

The typical shape of *Campylobacter* looks more like a spiral or helical one rather than a curved rod shape, which can change its shape into filamentous or coccoid to adapt to the stressful conditions [9, 10]. When two or more bacterial cells are grouped together, an "S" or a "V" gull-wing shape is formed [11]. The majority of *Campylobacter* species have a characteristic corkscrew-like motion due to a single polar flagellum at one or both ends of the cell [11], with the exceptions that *C. gracilis* is non-motile and *C. showae* contains multiple flagella [12].

Escherich observed the *Campylobacter*-like non-culturable spiral-shaped organism in infants' stool samples in 1886 [1, 13]. These bacteria, called related *Vibrio* by then, were first isolated from the uterine mucus of a pregnant sheep from a flock of 150 ewes that were experiencing an abortion rate of 33% in 1906 by McFadyean and Stockmanin in the United Kingdom [14]. A few years later, an apparently identical organism, firstly named as a spirillum and then as *Vibrio fetus*, was isolated from the fetal membranes of aborting cattle by Smith and Taylor in the United States [15, 16]. In 1949, Stegenga and Terpstra demonstrated the pathogenic role of *V. fetus venerealis* in enzootic sterility in cows [1, 17]. In 1931, winter dysentery in calves was attributed to infection with a "*vibrio*" that they called *Vibrio jejuni* [18], and a similar organism associated with swine dysentery was identified by Doyle [19].

Campylobacter-like organisms, were first isolated from humans in 1938 from the blood of patients suffering from diarrhea in a milk-borne outbreak affecting 355 people in the United States, and the causative bacterium for this outbreak was named "*V. jejuni*" [20]. This has been regarded as the first well-documented instance of human *Campylobacter* infection [1]. Subsequently, *V. fetus* was isolated from the blood of three pregnant women admitted to hospital because of fever of unknown origin in 1947 [21], and King described the isolation of "related *Vibrio*" from blood samples of children with diarrhea in 1957 [22]. Up to 1972, only 12 cases of 'related vibrio'infections were reported in the literature [1]. The reason for these small numbers of the reports was that the optimal or specific selective culture techniques for the isolation of 'related vibrio', later called as *Campylobacter* [23, 24], from feces were not developed at that time.

The first successful isolation of *Campylobacter* from feces or stool samples of patients with diarrhea was accomplished in Belgium using the technique of direct membrane filtration onto agar medium containing several antibiotics in 1968 and published in 1972 [25]. This study also found that *C. jejuni* was highly susceptible to erythromycin [26]. Consequently, in 1977, a selective culture procedure was recommended using selective medium containing antibiotics with incubation at 43°C in a microaerobic atmosphere (5% oxygen, 10% carbon dioxide, and 85% hydrogen) [27]. This facilitated the isolation of campylobacters, *C. jejuni* and *C. coli* (the two species known at the time to cause gastroenteritis), with greater ease [27]. The successful isolation of campylobacters from human feces based on the above selective media led to the recognition that *Campylobacter* is a leading cause of human diarrheal illness in many countries [5, 28].

Since the establishment of genus *Campylobacter* in 1963 with *Campylobacter fetus* as the species type [23], the taxonomy of the family *Campylobacteraceae* has transformed extensively [12]. The genus *Campylobacter* belongs to the family *Campylobacteraceae* proposed in 1991, the order *Campylobacterales*, the class *Epsilonproteobacteria*, and the phylum *Proteobacteria*. The class *Epsilonproteobacteria* presently comprises four closely related genera, *Campylobacter*, Arcobacter, *Dehalospirillum* and *Sulfurospirillum* [http://www.bacterio.net/index.html]. By 2021, the genus *Campylobacter* consists of 39 validly published species, 11 subspecies and 4 biovars [https://lpsn.dsmz.de/genus/campylobacter] (**Table 1**).

In addition, species within the *Campylobacter* genus can be grouped according to association with different host environments (e.g. animal intestinal tracts and

Taxon	Host	Disease		Foodborne	References
		In human	In animal		
C. armoricus	Surface water, human stool	humans displaying enteric infection	Unknown		[29]
C. aviculae	Lab Zebra Finches	Unknown	Unknown		[30]
C. avium	Poultry	Unknown	Unknown	Yes	[31, 32]!!!
C. blaseri sp. nov.	Common seals	Unknown	Unknown		
C. canadensis	Whooping cranes	Unknown	Unknown	Yes	
C. coli	Pigs, poultry, ostriches, cattle, sheep, penguin	Gastroenteritis meningitis, acute cholecystitis	Gastroenteritis, infectious Hepatitis	Yes	
C. concisus	Humans, domestic pets	Gastroenteritis, periodontitis, IBD	Unknown	Yes	
C. corcagiensis	lion-tailed macaques	Unknown			
C. cuniculorum	Rabbits	Unknown	Unknown		
C. curvus	Humans	Periodontitis, gastroenteritis	Unknown		
C. estrildidarum	Lab Zebra Finches	Unknown	Unknown		[30]
C. fetus subsp. *fetus*	Cattle, sheep, reptiles	Gastroenteritis, septicemia	Spontaneous abortion		
C. fetus subsp. *venerealis*	Cattle, sheep	Septicemia	Infectious infertility		
C. fetus subsp. *venerealis* bv. *intermedius*	Cattle	Unknown	Genital campylobacteriosis		
C. fetus subsp. *testudinum*	Human, reptile	Unknown	Unknown		
C. geochelonis sp. nov	Western Hermann's tortoise	Not known	Unknown		
C gracilis	Humans	Periodontitis	Unknown		
C. helveticus	Dogs, cats	Periodontitis	Gastroenteritis		
C. hepaticus sp. nov	Poultry	Unknown	Sporty liver disease		[33]
C. hominis	Humans	Unknown	Unknown		
C. hyointestinalis subsp. *hyointestinalis*	Cattle, deer, pigs, hamsters	Gastroenteritis	Gastroenteritis	Yes	
C. hyointestinalis subsp. *lawsonii*	Pigs	Unknown	Unknown	Yes	
C. insulaenigrae	Seals, porpoises	Unknown	Unknown	Yes	
C. iguaniorum sp. nov	Reptile (lizards)	Unknown	Unknown		
C. jejuni subsp. *doylei*	Humans	Septicemia, gastroenteritis	Unknown	Yes	
C. jejuni subsp. *jejuni*	Poultry, cattle, pigs, ostriches, wild birds, penguin	Gastroenteritis, Guillain-Barré syndrome	Spontaneous abortion, avian hepatitis	Yes	

Taxon	Host	Disease		Foodborne	References
		In human	In animal		
C. lanienae	Cattle	Unknown	Unknown	Yes	
C. lari subsp. *concheus*	Shellfish	Gastroenteritis	Unknown	Yes	
C. lari subsp. *lari*	Wild birds, dogs, poultry, shellfish, horses	Gastroenteritis, septicemia	Avian gastroenteritis	Yes	
C. lari subsp. *ornithocola*	Wild bird	Unknown			
C. mucosalis	Pigs	Unknown	Unknown	Yes	
C. novaezeelandiae sp. nov.	Birds and water	Unknown	Unknown		
C. peloridis	Shellfish	Unknown	Unknown	Yes	
C. portucalensis	Preputial mucosa of bulls	Unknown	Unknown		[34]
C. rectus	Humans	Periodontitis	Unknown		
C. showae	Humans	Periodontitis	Unknown		
C. sputorum bv. *sputorum*	Humans, cattle, pigs, sheep	Gastroenteritis, abscesses	Spontaneous abortion	Yes	
C. sputorum bv. *fecalis*	Sheep, cattle	Unknown	Unknown		
C. sputorum bv. *paraureolyticus*	Human, cattle	Unknown	Unknown		
C. subantarcticus	Birds in the subantarctic	Unknown	Unknown	Yes	
C. taeniopygiae	Lab zebra finch	Unknown	Unknown		[30]
C. troglodytes sp. nov	Wild chimpanzees	Unknown	Unknown		
C. upsaliensis	Dogs, cats	Gastroenteritis	Gastroenteritis		
C. ureolyticus	Humans	Gastroenteritis, Crohn's disease	Unknown		
C. volucris	Black-headed gulls	Unknown	Unknown		
C. vulpis	wild red foxes fox	Unknown	Unknown		

!: Valid species (39 species, 11 subspecies (subsp), and 4 biovars (bv)) as included on the website: https://lpsn.dsmz.de/genus/campylobacter
!!: Indicated as "sp. nov" in the table.
!!!: All the reference is referred to [31] and [32], except those indicated differently.

Table 1.
List of validly published! And newly proposed!! Species, subspecies and biovars in the genus campylobacter *and their common hosts and disease associations in humans and animals.*

human oral cavity) [35] and propensity to cause disease in animal and human hosts [13, 36] and (**Table 1**). Zoonotic *Campylobacter* species are commensal organisms found in the intestinal tract of a variety of mammals, birds and reptiles as well as in related environments, including water and soil [8, 11, 13, 37]. Among the zoonotic species, *C. jejuni* is responsible for approximately 81% of human gastrointestinal-related *Campylobacter* infections with *C. coli*, *C. fetus*, *C. lari* and *C. upsaliensis* being responsible for 8.4%, 0.2%, 0.1% and 0.09%, respectively. The remainder of human *Campylobacter* infections are from other species or undifferentiated "*campylobacters*" [38]. Other zoonotic species including *C. fetus*, *C. hyointestinalis*, *C. upsaliensis, C. sputorum, C. concisus* and *C. ureolyticus* have also been recognized as causal agents of human gastroenteritis [2, 37]. Of these, *C. concisus*, *C. upsaliensis* and *C. ureolyticus* are considered as emerging or under-recognized disease-associated species, where due to the advances in molecular biology and culture methodologies these species are becoming increasingly recognized as pathogens [7, 11]. In addition to causing gastroenteritis in humans, a number of *Campylobacter* species are potential oral pathogens and have been commonly isolated from the human oral cavity. These include *C. concisus*, *C. showae*, *C. gracilis*, *C. curvus*, *C. rectus*, and *C. ureolyticus* [35]. While *campylobacters* do not typically cause disease in animals, *C. fetus* subspecies *fetus* and *C. fetus* subspecies *venerealis* are important causes of reproductive system disorders and abortions in ruminants, and particularly *C. fetus* subsp. *venerealis* is restricted to cattle and causes bovine genital campylobacteriosis, and *C. hepaticus* causes Spotty Liver Disease in layer chickens [13, 33, 36] (**Table 1**). Based on the route of transmission to humans, campylobacters can also be classified as foodborne or non-foodborne groups (**Table 1**) [31]. This chapter will focus on the foodborne campylobacters potentially from food and food-related environments.

King [22] observed that the incubation at 42°C enhanced the growth of campylobacters, which led to the concept of "thermophilic" campylobacters. Traditionally, based on the tolerance to the temperature 42°C - 43°C (able to grow at this temperature), campylobacters are also referred as "thermophilic" and "non-thermophilic" groups, with *C. jejuni*, *C. coli*, *C. lari* and *C. upsaliensis* as the four human pathogenic species of *Campylobacter* and are often referred to as the thermophilic *Campylobacter* species. However, these species do not exhibit true thermophily (growth at 55°C or above) [39]. For example, the D value (the time it takes to reduce a microbial population by 1 logarithm) at 50°C for *C. jejuni* strains in skimmed milk was demonstrated to be between 1.3–4.5 min, and inoculation of a heat tolerant strain of *C. jejuni* into roast beef at a level of 5.9×10^6 /g resulted in no survivors by the time the internal temperature had risen to 55°C [40]. Similarly, a low D value between 0.7–1.4 at 60°C for *C. jejuni* and *C. coli* strains was found [41]. Therefore, it was suggested that the organisms which could grow at 41–43°C, should be referred to as "thermotolerant" *Campylobacter* species [42]. Therefore, the term "thermotolerant" is used in this chapter.

1.2 Distribution and diseases

Campylobacter species are commensal microorganisms of the gastrointestinal tract of many wild animals (birds such as ducks and gulls), farm animals (cattle and pigs), and companion animals (such as dogs and cats) (**Table 1**) [8, 11, 13]. The organisms can also be found in the internal organs of animals [43, 44]. The main route of transmission to humans is generally believed to be foodborne, via undercooked meat and meat products, particularly the poultry products, as well as raw or contaminated milk [11, 13]. Despite it is well-known fastidious nature, *Campylobacter* is also isolated from environmental sources, such as lake, river, soil, sea, and sewage, suggesting that environmental water is a possible vehicle that transmits *Campylobacter* to humans [8]. Different species of the *Campylobacter* genus naturally colonize a wide range of hosts (including pets, farm animals and wild animals) and are frequently detected in contaminated food products, indicating that these organisms are potentially transmissible to humans [7].

Many *Campylobacter* species are known pathogens in either humans or animals, or both [7] (**Table 1**). The main disease in humans caused by campylobacters is gastroenteritis worldwide, which is mainly due to *C. jejuni* [2, 7] and accounts annually for approximately 166 million foodborne illnesses around the world [5]. Meanwhile, *C. jejuni* infection may lead to autoimmune conditions known as Guillain-Barré syndrome (GBS) and Miller Fisher syndrome. *Campylobacter* species have also been associated with a range of gastrointestinal conditions, including inflammatory bowel diseases (IBD), Barrett's esophagus, and colorectal cancer [7]. In addition, they have been reported to be involved in extragastrointestinal manifestations, including bacteremia, lung infections, brain abscesses, meningitis, and reactive arthritis, in individual cases and small cohorts of patients [2, 7]. The precise role of *Campylobacter* species in the development of these clinical conditions is largely unknown. A growing number of *Campylobacter* species other than *C. jejuni* and *C. coli* have been recognized as emerging human and animal pathogens. The development of new molecular and innovative culture methodologies enhanced the detection and isolation of a range of under-recognized or emerging, and nutritionally fastidious *Campylobacter* species, including *C. concisus*, *C. upsaliensis* and *C. ureolyticus*. It has been found that these emerging *Campylobacter* species are associated with a range of gastrointestinal diseases, particularly gastroenteritis, IBD and periodontitis. Some cases of the gastrointestinal tract infection by these bacteria can lead to life-threatening extragastrointestinal diseases [2, 7]. Based on the high number of recent *Campylobacter* infections reported in the region, campylobacteriosis ranks as the third cause of death behind listeriosis and salmonellosis in the EU [38].

1.3 Objectives

Various methods have been used for detecting campylobacters including conventional culture methods, molecular (such as polymerase chain reaction or PCR) and immunological methods, and genome sequencing analysis. Currently, isolation and subsequent identification of the target campylobacters are required by most of the regulatory bodies globally. The multiple species of *Campylobacter* exhibit diverse physiological and metabolic characteristics, and growth requirements, which can interfere with the sensitivity and specificity of culture-dependent methods. Furthermore, strains among each species may behave differently in various culture media and under various culture conditions. Therefore, it is important to apply appropriate isolation and identification methods for different types of species and samples based on specific purposes. The subsequent detection and characterization can also be challenging for comprehensive and accurate identification, particularly for source attribution and epidemiological investigations. This chapter will firstly reviews the physiology of the organism in order to better understand the isolation procedure, followed by a review of the culture-dependent detection methods in combination with technologies newly developed for various *Campylobacter* species from food and food-related environment that may contaminate the food chain.

2. Physiology of campylobacters and growth requirements of different species

It is important to understand the unusual physiology of campylobacters *in vivo* and *in vitro* for the development of optimal culture procedures *in vitro*. A review by Park [45] indicated that compared with most foodborne bacterial pathogens which are considered to have stronger ability to survive the harmful conditions imposed by food processing and preservation, *Campylobacter* species require uniquely fastidious growth conditions and are usually more sensitive to environmental stress. It is generally believed that *campylobacters* may also lack many of the well-characterized adaptive responses that support the resistance to stress in other bacteria [45]. These factors may lead to the difficulty to recover the campylobacters from food and food-related environment. However, several reports indicated that campylobacters have peculiar mechanisms such as their inherent genome plasticity and gene regulators that respond to changing environments which enhance their survival in hostile conditions [46–48].

2.1 Microaerobic requirement

As microaerophilic and capnophilic microorganisms, *Campylobacter* species generally require a microaerophilic atmosphere with reduced oxygen (approximately 5–10% O_2) and elevated carbon dioxide (5–10% CO_2) concentrations for its optimal growth *in vitro* [2, 11]. They have a respiratory type of metabolism. Several *Campylobacter* species including *C. concisus, C. curvus, C. gracilis, C. mucosalis, C. rectus, C. showae* and some strains of *C. hyointestinalis* require extra hydrogen (3–7% H_2) or formate acting as an electron donor for microaerobic growth and successful recovery. Although most of the *Campylobacter* species require microaerobic conditions for growth, however certain species such as *C. concisus* can grow under or prefer anaerobic conditions for growth [2, 49]. There are also some strains being aerotolerant [50–52].

2.2 Optimal growth and survival temperature, pH and water content

The minimum, optimum and maximum growth temperatures (°C) of campylobacters are 32, 42–43 and 45 respectively, and minimum, optimum and maximum growth pH values are pH 4.9, 6.5–7.5 and 9.5 [11, 45, 53]. *Campylobacter* species will not survive below a pH of 4.9 and above pH 9.0 and grow optimally at pH 6.5–7.5 [11]. The organisms will grow with water activity (a_w) at 0.987 and 0.997 [11, 45, 53]. Growth does not occur in environments with *a*w lower than 0.987 (sensitive to concentrations of sodium chloride (NaCl) greater than 2%w/v), while optimal growth occurs at $a_w = 0.997$ (approximately 0.5% w/v NaCl) [11].

Generally, some *Campylobacter* species (e.g. *C. jejuni, C. coli, C. lari, C. upsaliensis, C. helveticus* and *C. insulaenigrae*) are referred to thermotolerant with an optimal growth temperature of 37–42°C and a maximum temperature of ~46°C. The remaining *Campylobacter* species are considered not thermotolerant in the literature, with an optimal growth temperature of 37°C [54, 55]. However, based on the official publications for the validated species and subspecies in *Campylobacter* genus classification up to date (https://lpsn.dsmz.de/genus/campylobacter; accessed on 2021-12-21), among 39 validly published species, 28 species with 2 subspecies are indicated as thermotolerant (able to grow at 42°C) with the majority of the strains (90–100% strains) able to grow at 42°C [**Table 1** with relevant references]. These thermotolerant species/subspecies include *C. armoricus*, *C. aviculae* sp. nov., *C. avium*, *C. blaseri*, *C. canadensis*, *C. coli*, *C. corcagiensis*, *C. estrildidarum* sp. nov., *C. helveticus*, *C. hepaticus*, *C. hyointestinalis* subsp. *hyointestinalis*, *C. hyointestinalis* subsp. *lawsonii*, *C. jejuni* subsp. *jejuni*, *C. lanienae*, *C. lari* subsp. *concheus*, *C. lari* subsp. *lari*, *C. mucosalis*, C. *novaezeelandiae*, *C. ornithocola*, *C. peloridis*, *C. portucalensis*, *C. sputorum*, *C. subantarcticus*, *C. taeniopygiae* sp. nov., *C. upsaliensis*, *C. volucris* and *C. vulpis*. Seven species with three subspecies are non-thermotolerant or poorly thermotolerant (only 0–10% strains thermotolerant), including *C. fetus* subsp. *venerealis*, *C. geochelonis*, *C. iguaniorum*, *C. insulaenigrae*, *C. jejuni subsp. doylei*, *C. pinnipediorum* subsp.

caledonicus and *C. pinnipediorum* subsp. *pinnipediorum*. Others are partially thermotolerant (26–89% strains), which can be divided into three groups, namely, (−) (11–25% strains are thermotolerant) including *C. hominis* and *C. rectus*; (V) (26–74%) including *C. curvus, C. fetus* subsp. *testudinum*, *C. gracilis*, *C. showae and C. ureolyticus*; and (+) (75–89%) including *C. concisus*, *C. cuniculorum* and *C. fetus* subsp. *fetus*. It is worthwhile to note that there are different degrees of tolerance to growth temperatures among the strains of the same species, which should be taken into consideration when the incubation temperatures are chosen for the diagnostic testing.

Generally speaking, campylobacters are unable to grow below 30°C [45, 56]. It was suggested that the absence of cold-shock proteins might be responsible for the inability of this pathogen to grow at lower temperatures [57]. However, although not growing, *C. jejuni* was found to survive for more than 4 h at 27°C and 60–62% relative humidity on food contact surfaces [58]. These physiological characteristics reduce the ability of campylobacters to multiply outside of an animal host, and in food during their processing and long term storage [45]. Due to the fluctuation of body temperatures in reptiles, *Campylobacter* species in reptiles have adapted to larger temperature ranges and are more tolerant to lower temperatures than those found in mammals and birds. For example, the proposed *Campylobacter geochelonis* sp. nov., isolated from western Hermann's tortoise grows at 25°C and not at 42°C [54].

In pure cultures, *Campylobacter* spp. are normally inactivated by frozen storage at −15°C in as few as 3 days [59], although freezing does not eliminate the pathogen from contaminated foods [60]. Hazeleger [57] revealed that aged *C. jejuni* cells survived the longest at 4°C. Under the cold and other stress conditions, campylobacters may enter viable but non-culturable (VBNC) state [11]. The campylobacters at VBNC state may affect the sensitivity of culture-dependent detection procedures.

2.2.1 Antibiotics resistance

Most campylobacters are resistant to a few antibiotics including amphotericin B, cefoperazone, colistin, cycloheximide, polymyxin B, rifampin and trimethoprim at different concentrations, which have been used as supplements in selective media [61]. In addition, *C. jejuni*, *C. coli*, *C hyointestinalis* and *C, fetus*, but not *C. upsaliensis*, have also been shown to be resistant to Aztreonam [62]. There is evidence that some strains of *C. coli* and even a few strains of *C. jejuni* are likely to have been missed due to their sensitivity to cephalothin. Several species, including *C. hyointestinalis*, *C. upsaliensis* and *C. fetus* are inhibited by the high amount of cefoperazone contained in the selective medium [61, 63, 64]. Generally speaking, *C. upsaliensis*, a commonly believed to be an important human pathogen, is sensitive to the antibiotics routinely used in *Campylobacter* selective media. The recovery of this organism will rely on the development of widely applicable, effective techniques for its isolation, such as membrane filtration based techniques [63]. The presence of these antibiotics in the selective media generally used for the isolation of *Campylobacter* species (e.g., Skirrow's medium) may well lead to the suboptimal identification of *C. upsaliensis* in clinical specimens at most centers. It has been shown that the procedure (Cape Town Protocol) using membrane filtration and antibiotic-free agar isolate more *Campylobacter* species than the agar alone with antibiotics [65]. Therefore, the usage of antibiotics must be carefully selected for the culture media to avoid the false negative results.

2.2.2 Essential nutrient requirements for growth in vitro

Campylobacters obtain their energy sources from amino acids or tricarboxylic acid cycle intermediates [6, 66]. *Campylobacter* is generally considered a nonsaccharolytic bacterium because of inability to ferment or use glucose and other carbohydrates as growth substrates, which has been supported by genome sequence analysis [67–69] and recent growth-independent phenotype microarray analyses that allows monitoring the respiratory activity of metabolically active cells [70]. Furhter studies confirmed that pentoses and hexoses like glucose, fructose, galactose, rhamnose and the disaccharides lactose, maltose, trehalose and sucrose do not enhance the respiratory activity of *C. jejuni* [66, 71–73]. The *Campylobacter* culture media contains mainly the peptamin or peptone that provides amino acids (source of carbon), sulfide, and nitrogen required for making their energies, yeast extract provides B vitamins (coenzymes), and sodium chloride to maintain osmotic equilibrium [74]. *Campylobacter* enrichment broth must contain sources of iron, such as blood, hemin, and ferrous sulfate [61].

2.2.3 Considerations for developing a new media or culture procedure or strategies

Taken together, the studies of metabolic activities have been mainly focused on *C. jejuni*, which showed an intriguing metabolic diversity among different strains. The diverse growth properties of *C. jejuni* isolates result from the presence or absence of various metabolic genes involved in the strain-specific utilization of particular substrates such as fucose, asparagine or glutamine and peptides. In addition, *C. jejuni* isolates contain different sets of group A chemoreceptor *tlp* genes that respond differently to potential nutrients [75, 76]. The presence of various chemosensory receptor genes in *C. jejuni* suggests that different strains may not respond equivalently to certain nutrients and consequently cannot utilize and ben efit from the same growth substrates. These may affect the isolation of all the strains from the same *Campylobacter* species using the same medium.

3. Isolation and detection

Traditionally, campylobacters have been detected using culture-dependent procedures followed by the identification, confirmation and typing of the isolates using biochemical, immunological and molecular methods [11, 13, 77]. Cultural isolation remains the gold standard for confirming the presence of live bacteria in a sample. Molecular and immunological methods, particularly PCR [13], and recently the whole genome sequencing tools [78–82] have also been used for detection of *Campylobacter* species in different sources. The following will review the methods of culture, identification and confirmation, and, characterization of the isolates in detail.

3.1 Culture methods

3.1.1 History of methods development

Campylobacter-like organisms were first isolated from the blood of humans in 1938 using tryptose phosphate beef broth or brain broth at 37°C, but cultures using media including Endo medium, blood agar, plain agar, Herrold's, Loeffler's, liver agar, and milk agar for fecal samples failed [1, 20]. King in 1957 [22] demonstrated that campylobacters referred to as the *V. fetus* group and the "related vibrios" by then, grew in thioglycollate medium under microaerophilic condition in a Brewer anaerobic jar. Some strains grew on MacConkey's agar, but failed to grow on several solid media such as SS (*Shigella-Salmonella*) agar, Simmons' citrate agar or Christensen's urea agar. In addition, it was found that the *V. fetus* strains preferred a temperature between 25°C and 37°C with very little or no growth occurring at 42°C, while "related vibrios" (now *C. jejuni*) strains failed to grow at 25°C, but grew optimally at 42°C, with no growth occurring at 52°C. The first successful isolation of *Campylobacter* from feces of patients was accomplished in 1968 and published in 1972, by using the technique of direct membrane filtration (0.65 µm pore size) onto fluid thioglycolate-agar medium containing antibiotics (bacitracin, polymyxin B sulfate, novobiocin and actidione [1, 25]. This was followed by the development of a selective medium in 1977, which enabled the isolation of *C. jejuni* and *C. coli* from human feces [27]. In Skirrow's study, initially the filtrates of fecal suspensions prepared by passing samples through Millipore filters based on that described by [25] were cultured on plain blood agar. Then this study also developed a procedure using medium containing vancomycin, polymyxin B, and trimethoprim, and incubation condition at 43°C in an atmosphere of 5% oxygen, 10% carbon dioxide, and 85% nitrogen [27]. The successful isolation of campylobacters from human feces based on the above selective media led to the recognition that *Campylobacter* is a leading cause of human diarrheal illness in many countries [5, 28]. Since then, extensive efforts have been made to develop, evaluate and validate the existing and new media, and procedures for the isolation and identification of campylobacters from food and food-related environments.

The main media and procedures currently in use were developed during 1980s [61]. The characteristics of these organisms allowing growth and selection have been largely understood [83]. The prescence of *Campylobacter* cells in larger numbers from clinical (or animal fecal) sample makes their recovery reasonably straightforward. The ability of thermotolerant campylobacters particularly *C. jejuni* and *C. coli*, to grow at 42°C, has enhanced the selective isolation of thermotolerant campylobacters by inhibiting the growth of other bacterial species. The applications of antibiotic cocktails in selective media have become refined in their combinations of the types and concentrations over time with continuously improved bacterial recovery rates. For food and water samples, and perhaps clinical samples requiring lengthy periods of transportation before analysis, the recovery of stressed cells can also be challenging [11, 83]. The procedures for the isolation of *Campylobacter* spp. from foods were adapted originally from clinical microbiology protocols since the 1970s [61, 84]. So far more than 20 of each selective enrichment broth and selective agar media have been developed (**Table 2**) with different scales of evaluation and validation [13, 37, 61, 77, 83, 85]. As campylobacters do not ferment carbohydrates, peptones are included in all media as a nutrient source. Some such as Preston broth, Bolton broth, Exeter broth and *Campylobacter* enrichment broth contain meat or yeast extract [61]. Most media are developed based on several commonly used media such as nutrient broth or agar, Brucella, Columbia, thioglycollate media with the addition of various antibiotics and with or without blood [61]. Most *Campylobacter* media contain antibiotics and blood which neutralizes trimethoprim antagonists. Oxygen quenching agents are also used to overcome the adverse effects of toxic oxygen derivatives that can form when media are exposed to light (e.g. hydrogen peroxide and superoxide) [61]. Selective or non-selective blood agar media were successful in isolating new *Campylobacter* species [61]. Depending on the purposes, quantitative and qualitative procedures have been developed. Currently there are a few agar plates available for detection and enumeration purposes. In addition to those developed in the early days including the most used currently, modified charcoal cefoperazonedeoxycholate agar (mCCDA), Skirrow, Campy BAP, Karmali, and Abeyta-Hunt-Bark, several new agar plates have been developed, including Campy-Line, Campy-Cefex and several commercially available chromogenic agars during the last two decades (**Table 2**) [96]. Currently, the most commonly used enrichment media include Bolton broth, Preston broth and Exeter broth [85].

Enrichment broth media		Agar media	
Name	**Reference**	**Name**	**Reference**
Thioglycollate broth	[22, 61]	Thioglycollate	[61, 85]
Preston broth	[61]	Dekeyser	[85]
VTP Brucella-FBP broth (VTP) or VTP FBP broth	[61]	Skirrow blood	[27, 61, 85]
Modified charcoal cefoperazone deoxycholate (mCCD) broth	[61]	Blaser A	[85]
Doyle and Roman enrichment broth	[61]	Blaser-Wang	[85]
Park and Sanders broth	[61]	Butzler and BU40 (modified Butzler) (contains bacitracin, novobiocin, cycloheximide, colistin, and cefazolin)	[61]
Exeter broth	[61]	Butzler (Virion) (BV) (cefoperazone, rifampin, colistin, and amphotericin B)	[61]
Christopher broth	[85]	Butzler (Oxoid)	[61]
Lander broth	[85]	Butzler selective medium or Campy BAP	[61]
Waterman broth	[85]	Preston agar (Campylobacter Agar)	[61]
Hunt and Radle broth	[61]	Waterman agar	[85]
Bolton broth (Campylobacter enrichment broth – Bolton formula)	[61]	Modified Butzler agar (MBA)	[61, 86]
Campylobacter enrichment broth (CEB) (a commercial version of Bolton broth with A commercially available enrichment broth varies only in the substitution of natamycin for cycloheximide	[87]	Blaser medium or Campy BAP (with vancomycin, trimethoprim, polymyxin B, cephalothin, and amphotericin)	[61]
Rosef and Kapperud Campylobacter enrichment broth (RKCEB),	[61, 86]	Charcoal cefoperazone deoxycholate (CCDA) (Campylobacter Blood-Free Selective Agar)	[61]
Rosef	[61]	mCCD agar	[61]
Lovett's broth (Brucella broth with FBP, vancomycin, polymyxin B)	[88]	Campylobacter Selective medium (CAT)	[61]
Blood-free Campylobacter medium (BFCM)	[86]	Karmali agar	[61]
Mueller and Hinton broth without or with antibiotics (MHBH)	[61]	Campy Brucella agar (CBAP),	[61, 86]
Fennell's medium	[61]	Mueller and Hinton agar with antibiotics (MHBA) or without antibiotics	[61]
Semi-solid medium or selective semisolid Brucella medium (SSBM),	[61, 86]	Columbia Blood Agar	[61]

Enrichment broth media		Agar media	
Name	**Reference**	**Name**	**Reference**
Buffered Peptone Water (transportation and carcass rinse and pre-enrichment)	[61, 89, 90]	Abeyta–Hunt–Bark (A-H-B) agar (Heart infusion agar with yeast extract and antibiotics)	[91]
Weybridge's (Transportation media)	[92]	Campy-Cefex	[93]
Wang's semi-solid Transportation Medium	[89]	Campy Line agar with sulfamethoxazole (CLA-S),	[94]
Wang's Freezing/storage Medium	[89]	CampyFood agar Internationally validated method for detection & enumeration (ISO 16140/AOAC)	BioMérieux, France
Cary-Blair transportation medium	[91]	CASA Chromogenic Medium for enteric *Campylobacter* species	AES Chemunex, France
A-H slant	[91]	RAPID' Campylobacter agar (chromogenic)	Bio-Rad (Certified NF VALIDATION according to the ISO 16140 standard)
Brain Heart Infusion broth (motility testing)	[95]	CHROMagar™ *Campylobacter* CAC (chromogenic for thermotolerant *Campylobacter*)	CHROMagar, Paris, France
The aztreonam amphotericin vancomycin (AAV) experimental campylobacter selective medium	[61, 62]	Brilliance *CampyCount AGAR* (chromogenic for *Campylobacter jejuni* and *Campylobacter coli*)	ThermoFisher Scientific
CampyFood broth (Internationally validated method for detection & enumeration (ISO 16140/AOAC))	BioMérieux, France	Campylosel	Bio-Mérieux, France

Table 2.
List of culture media used for isolation, transportation and maintenance of campylobacters.

3.1.2 Current culture procedures

So far, three types of basic culture procedures have been commonly employed for the isolation of campylobacters. These methods include 1) membrane filtration onto non-selective or selective agar media; 2) direct plating on selective agar, either blood-based or charcoal-based; and 3) selective enrichment in broth followed by streaking onto selective agar [61, 77]. Various culture supplements and procedures have been examined or standardized to improve selective isolation of *Campylobacter* species [11, 13, 37, 61, 77, 83]. The above three approaches have been used for qualitative detection or enumeration (semi-quantitative and quantitative) using the most probable number (MPN) procedure or direct plating method [97].

Although many culture-independent detection methods have emerged over time, isolation by culture is still the "gold" standard procedure for the detection of campylobacters for regulatory bodies. Despite the continued improvements in the isolation procedures of *Campylobacter*, however, challenges remain, which reduce the efficiency of these methodologies [98]. As discussed in the section related to the growth requirements, *Campylobacter* exhibits dynamic and malleable physiological and metabolic characteristics that have impacts on the sensitivity and specificity of culture-dependent methods. So far, there is no single or "standard" accepted method of isolating and detecting all *Campylobacter* species from all kinds of sample types, due to different requirements of temperature, microaerobic conditions, nutrients and susceptibility to selective antibiotics. However, there are some generally agreed procedures for common types of *Campylobacter* species and some types of food samples [13, 37, 61, 77, 83, 85]. Several protocols have been published or recommended by recognized authorities, such as 1) the International Standards Organization [99, 100]; 2) US Food and Drug Administration (FDA) [91]; 3) the U.S. Department of Agriculture (USDA) - Food Safety and Inspection Service (FSIS) [89]; 4) the Public Health England [95]; 5) World Organization for Animal Health (OIE) [92]; 6) Health Canada (HC) [101]; 7) Australia and New Zealand [102]. These methods use the most effective protocols to isolate thermotolerant *Campylobacter* spp. (mainly *C. jejuni* and *C. coli*) from food, primarily poultry products [77]. Cape Town protocol is also a well-known protocol used in South Africa, which employs membrane filtration and antibiotic-free agar plating for a broad spectrum of *Campylobacter* species [65].

The method of choice to isolate low numbers of *Campylobacter* from contaminated food samples is the combination of enrichment broth with selective plating or direct plating on selective agars. For analysis of fecal samples or certain types of food samples, direct plating is often preferred due to the presence of large numbers of non-stressed campylobacters in feces. Several selective agars have been used for various purposes including regulatory requirement in US [65, 89, 95]. However, due to the slow growth of *Campylobacter* species, many are lost to competition by contaminant bacteria naturally present in foods. As mentioned earlier, certain antibiotics in the selective media may inhibit the growth of certain *Campylobacter* species and strains. By taking advantage of the unique motility of campylobacteria, the membrane filtration method known as "Cape Town protocol" allows these organisms in samples to penetrate the cellulose filters of 0.45-mm or 0.65-mm pore sizes to antibiotic-free blood or other agar. This method has shown a great advantage in isolating a wider spectrum of *Campylobacter* species, including *C. upsaliensis*, *C. concisus*, *C. curvus*, *C. rectus*, *C. sputorum* biovar *sputorum* and *C. hyointestinalis* as well as the standard *C. jejuni* and *C. coli* from human stools [65]. However, this method takes longer time, i.e., a couple of days more than conventional direct plating [103] and also depends on the presence of a large number of *Campylobacter* cells with cell motility [104]. The membrane filtration method has been evaluated for food and water samples [104–107]. One study [104] showed that the minimum numbers of motile bacteria required for this method were 2.2 and 2.1 log colony forming unit (CFU) for 24-h cultures and centrifuged cells, respectively, and 4.1 and 3.4 log CFU of coccoid and nonmotile mutant cells, respectively. Broiler meat samples after enrichment in Bolton's broth showed that approximately 1.7 log CFU of *Campylobacter* can be detected with pure colonies on agar plates using this filtration method. The results from the studies [104–107] demonstrate that the motility of the bacteria influences passage through cellulose filters and that 0.65-mm-pore-size filters on agar plates help obtain pure *Campylobacter* colonies from enriched food samples [104, 105]. A novel and simple filtration procedure after enrichment in Rosef's enrichment broth was developed using a hydrophobic grid membrane filter (HGMF) on antibiotic-free semisolid medium (SSM). The HGMF-SSM method showed higher recovery rates using turkey samples and pig fecal samples compared with Rosef's broth enrichment procedure [105]. A study using selective enrichment combined with membrane filtration has shown a similar recovery rate but with 20-fold fewer false-positive for campylobacters in water compared with the enrichment procedure [106]. Another study [107] showed that the filtration method, and real-time PCR and digital PCR were more sensitive than enrichment culture method using inoculated sprouts samples. The filtration method showed a similar detection ability to PCR in all samples.

For the enrichment procedures, particularly for the samples with low numbers of cells under stress, a preliminary (resuscitation) period of incubation at reduced temperature (37°C) for about 4 h prior to increasing the temperature to 42°C (for thermotolerant species) for the remainder of the 48 h of incubation time has been used commonly. However, procedures using staged periodic increases in temperature to aid adaptation and recovery can be time-consuming and overly complex [61]. Several factors have been considered during the development or application of the media for campylobacters from food, including incubation at 37°C instead of 42 or 43°C and changes in the types and concentrations of antibiotics or various combinations of selective enrichment broth with selective agars, in order not to inhibit a wider spectrum of the organisms such as *C. upsaliensis*, *C. jejuni* subsp. *doylei* and some strains of *C. coli* and *C. lari* [108–111]. The use of immunomagnetic separation methods to concentrate all cells and to remove competitive microorganisms could be used as an option. However, this method is problematic due to low capture efficacy because of the unique movement of campylobacters and with the few studies undertaken on campylobacters showing limited efficacy when applied to naturally contaminated samples [112], and possible surface antigenic variations of the campylobacters if a single antibody is used for capture [83]. Considerably less studied are the prevalence and importance of species other than *C. jejuni* and *C. coli*, especially as related to food as a source of illness. Most *Campylobacter* species have different growth requirements to *C. jejuni* and *C. coli*, and until recently, specific methods for isolation have not been applied. In order to enhance the detection sensitivity and accuracy, and shorten the turnaround time of culture process, the screening of the presence of campylobacters in the enrichment broth during or after enrichment could be conducted using various methods, such as PCR [89].

Microaerobic systems are also important to support *Campylobacter* growth *in vitro*. Different methods have been developed to generate microaerobic atmospheres for routine use during the enrichment of food samples or during the incubation of inoculated plate media. The microaerobic atmosphere is usually generated in a gassed jar system, either by continuous flow of the mixed gas through the containers or by evacuation and gas replacement. If a large number of samples are processed, the evacuation-replacement is a more economical and practical way, for which, the air in the jar or gas tank is removed by a vacuum pump, and then replaced with a desired microaerobic gas mix [61, 91, 113]. Other commercial sachets that generate microaerobic conditions can also be used, particularly for a small number of samples or when other evacuation-replacement system(s) not available [114]. Plastic bags utilized to freeze food products with a "ziplock" type closing to prevent air leaks have been successfully used with gas-generating sachets and manual evacuation-replacement systems to be flushed with a desired microaerophilic gas mixture [115, 116]. Due to specific growth requirements, certain species or stains require H_2 content [6]. For a large number of samples, or to create unique microaerobic gas mixes with increased H_2 content, more sophisticated microaerobic workstations can be used [44]. In addition to generating the microaerobic conditions, there have been several attempts to use O_2-quenching agents added to enrichment broths and agar plates for the isolation of *Campylobacter* species to reduce the toxic effects of oxygen radicals. These O_2-quenching agents include blood or alkaline hematin, charcoal, iron salts, norepinephrine, ferrous sulfate, sodium metabisulfite and sodium pyruvate (known as FBP supplement). In general, if blood or charcoal is added to agar plates, no other O_2 quenching compounds are required [61, 108]. Optimum growth can also be maintained in a tri-gas incubator [117] or a continued culture bioreactors [118]. Possibly even further neglected is the requirement for gaseous hydrogen in the cultivation atmosphere, without which many species cannot grow. From that perspective, the lack of commercially available"gas packs"to generate a microaerobic atmosphere that includes H_2 is unfortunate, especially since the growth of *C. jejuni* and *C. coli* are also enhanced in the presence of H_2. The inclusion of oxygen quenching supplements in pre-enrichment media seems to be a widely adopted practice to allow broth cultures to be incubated in air [61, 91, 101].

3.1.3 Regulatory use for risk assessment and control

The prevention of transmission to humans is paramount in reducing the incidence and burden of *Campylobacter* disease in humans. In the last two decades, extensive risk assessment and baseline studies on the distribution of the organisms in the food chain have been conducted in several countries [8, 31, 119]. The ubiquity of *Campylobacter* in the environment and poultry products presents difficulties in investigating the complex pathways for infection with no single specific point for effective prevention and control. Regardless of the high levels of contamination by this pathogen, particularly in poultry carcasses and its products, raw meat contaminated with *Campylobacter* is still allowed to be sold at retail in most countries. Since raw poultry is the main source of infection, poultry has been the focus for reduction with various success using different approaches in several countries including Iceland, New Zealand and UK [8, 31, 119]. Recently, the United States of America has introduced a standard and a compliance guide for poultry industries to reduce the campylobacters in raw poultry [120]. In Europe, a Process Hygiene Criterion (PHC) (Commission Regulation (EU) 2017/1495 of 23 August 2017 amending Regulation (EC) No 2073/2005) for *Campylobacter* spp. came into effect in January 2018. This PHC set a limit of the *Campylobacter* load (less than 1000 CFU/g) on broiler carcasses to control contamination of carcasses during the slaughtering process through monitoring and taking corrective actions when the mandated targets are breached [121].

3.1.4 Impacts of sampling and culture procedures on the recovery sensitivity and isolation of strain types

3.1.4.1 Detection sensitivity

According to the US Centers for Disease Control and Prevention (CDC), in approximately 80 and 56% of the cases of foodborne illness and death respectively in the USA, causal agents were not identified [122, 123]. In Canada, among 115 foodborne outbreaks reported from 2008 to 2014, 7.8% of outbreaks did not identify the etiologic agent [124]. Campylobacterosis is ranked as one of the major foodborne illnesses [123]. *Campylobacter* species. may account for some of the illnesses for which etiological agents were not identified. In fact, a study [125] in Canada indicated that current methods for isolation of *Campylobacter* species from clinical samples might fail to recover isolates from positive samples, particularly those in cryptic taxa of *Campylobacter*. Furthermore, patients may also be colonized with more than one genotype of *Campylobacter* [126]. Existing isolation methods have technical limitations in isolating this fastidious bacterium, such as a growth competition with indigenous bacteria in food samples. When compared with PCR methods, it was demonstrated that *Campylobacter* culture failed to correctly detect *Campylobacter* in 30% of positive patient stool specimens [127] or detected fewer species [128]. The studies comparing the culture and PCR methods for food have demonstrated that PCR methods identified more positive samples than culture methods [107, 129].

Several reasons could attribute to the relatively low sensitivity of culture methods compared with molecular methods. Sample collection and preparation for the sensitive recovery of live campylobacters may be important [77, 130]. For food and water samples, and perhaps clinical samples requiring lengthy periods of transportation before analysis, the recovery of stressed cells can be challenging [83]. Another interfering factor is the influence of microbiologically diverse and complex food or other sample matrices. A study [111] compared the effects of Bolton and Preston selective media on the microbiota compositions and isolation frequencies using next-generation sequencing (NGS) analysis of 16S rRNA. The results showed that Bolton and Preston-selective enrichments generated different microbiota communities and that the sequence of combining the selective media also critically affects the isolation frequency by altering microbiota compositions. For example, the highly prevalent *Escherichia coli* in Bolton media negatively affected the efficacy of *Campylobacter* isolation. The study [78] compared five different commercialized selective *Campylobacter* media for the ability to isolate *Campylobacter* from broiler fecal samples using 16S rRNA tagged-pyrosequencing of the isolated colonies. Sequencing results indicated that 0.04% of the total fecal microbial community was *Campylobacter*, and 88–97% of the putative colonies were in fact *Campylobacter*. The study also revealed that incubation atmosphere had little effect on recovery, but a significant difference in media specificity was found at 42 vs. 37°C. Different culture media showed different non-*Campylobacter* sequence types. Therefore, there are significant challenges in culturing *Campylobacter* on selective and/or differential media due to the presence of other competitive microorganisms, which can likely influence the metabolism of *Campylobacter*. In addition, the diversity of various host or environment matrices, such as poultry- or bovine-specific matrices, may also induce biochemical changes in *Campylobacter*, which further obscure isolation and identification. In addition, exposure to environmental stresses such as temperature, pH, aw, and starvation triggers a response that often results in the so-called "viable but non-culturable" (VBNC) form that appears to be capable of surviving as an intact and potentially infectious agent yet resistant to conventional culture [48, 83].

3.1.4.2 Impact on the isolation of different strain types

Most official protocols including that of the ISO, US FDA and HC require 25 g of meat for testing. However, retail packages typically contain multiple meat pieces with weights exceeding the required amount. This raises a concern that the testing of the required amount taken from only one of multiple pieces of meat from the same package may not be representative of the whole package. In addition, several studies have shown multiple strains could be present in a single chicken flock [131–134], chickens from a positive flock could contaminate the raw chicken meat at the slaughterhouse [134, 135] and internal organs and raw meat samples had multiple *Campylobacter* strains [43, 44]. Therefore, slaughter and packaging processes could lead to the contamination of chicken carcasses/raw meat from the same or different flock(s) with multiple strains. The selection of samples from the same package may affect the isolation of different strains in contaminated samples. Furthermore, the effects of various culture procedures on the isolation of different genotypes of campylobacters are not well understood. Overall, the culture conditions including temperature and types of media affected the genotypes of campylobacters recovered from raw chicken [136–138].

Taken together, specific isolation procedures and culture media influence the diversity of *Campylobacter* species recovered from samples. Temperature, media, time, and enrichment all influence the ability to isolate *Campylobacter.* The infective dose of the pathogen is low (approximately 500 CFU) [139, 140]. These studies highlight the importance of method sensitivity, and the need to collect multiple isolates from both clinical samples and potential sources of infection to support source attribution and epidemiological investigations.

3.2 Identification, confirmation and typing methods

The next step for the successful isolation of a strain is the identification and confirmation of the presumptive isolate. Traditionally, identification and confirmation of an isolate consisted of the examination of colony morphology, phase-contrast microscopic examination of morphology and corkscrew-like motility of the suspect, followed by confirmation using immunological assays (e.g., agglutination tests), biochemical and phenotypical testing, and molecular approaches [91, 101]. More recently, microbial identification based on proteomic profiling using Bruker ® MALDI Biotyper and genomic sequencing approaches have been recommended [82, 89, 98]. However, different published official protocols use different tests screening, confirmation and detection using agar colonies or enrichment broth. The identification and selection of suspect colonies from the selective agar are the first and crucial steps. To assist the accurate identification and selection of the *Campylobacter*-like colonies on the agar plate, particularly when the colonies are atypical or mixed with other floral microorganisms, specific, rapid and quantitative colony blot immunoassay [137, 141] and molecular tests, such as PCR, can be used [107].

The final stage in strain characterization is subtyping to allow rigorous assessment for epidemiology and source attribution purposes. Various strain subtyping approaches, including phenotyping and genotyping, have been developed and applied. The classic techniques for differentiating isolates phenotypically are based on the presence or absence of biological or metabolic activities expressed by the organism. Since 1980s, a few phenotyping schemes have been developed, including Skirrow-Benjamin and Preston biotyping schemes [142, 143], Penner (haemagglutination) and Lior slide agglutination serotyping schemes [144, 145], plasmid typing [146, 147], bacteriophage typing [148], and multilocus enzyme electrophoresis [149]. The most popularly used phenotypic methods to differentiate thermotolerant *Campylobacter* (mainly *C. jejuni* and *C. coli*) isolates include Skirrow-Benjamin biotyping scheme, Penner and Lior serotyping schemes, and multilocus enzyme electrophoresis [91, 101, 119, 149]. Although most of these methods lack discriminatory power, they are still used and are efficient to characterize bacterial food-borne pathogens [149]. All of the tests described above can be used alone or in combination to isolate and identify *Campylobacter* species, particularly *C. jejuni* and *C. coli*.

The need to enhance the limited discriminatory power of the traditional phenotyping methods in epidemiological investigations has led to the development of molecular typing methodologies. These improved technologies have been instrumental in reporting source attributions of sporadic infections and outbreaks with *Campylobacter* by providing information on the genetic subtypes. Many molecular subtyping methods have been developed to characterize *Campylobacter* species, but only a few are commonly used in molecular epidemiology studies [83, 138, 150, 151]. The commonly used methods include: pulsed-field gel electrophoresis (PFGE), *flaA* short variable region sequence typing (flaA-SVR), *flaA* restriction fragment length polymorphism analysis (*flaA*-RFLP), multi-locus sequence typing (MLST), extended MLST (eMLST), ribotyping, random amplification of polymorphic DNA (RAPD), microarray comparative genomic hybridization (MCGH), comparative genomic fingerprinting, single nucleotide polymorphisms, high-resolution melting analysis (HRM), Matrix-assisted laser desorption ionization-time of flight mass spectrometry (MALDI-TOF MS), and nucleotide and whole genome sequencing [150, 152]. The molecular typing methods have played a significant role in tracking sources of *Campylobacter* spp. infection [150]. However, most of the above-mentioned technologies are based only on a small fraction of the genome [150, 152]. In recent years, various emerging next generation sequencing (NGS) platforms and various pipelines for different purposes have been developed [153]. The NGS has been applied in the development of molecular tests, and in recent years various platforms have been used for whole strain genome sequencing for campylobacters for various purposes including strain differentiation with more discrimination power [82, 152, 154]. The method has been recommended for use with the detection and typing of campylobacters by several regulatory organizations such as USDA [89].

Overall, typing methods play an instrumental role in the identification, monitoring, and prevention of *Campylobacter* infections. The use of multiple phenotypic and genotypic or molecular typing methods can improve species and subspecies discrimination and is appropriate when trying to identify pathogenic organisms like *C. jejuni, C. coli,* and *C. laridis* [151]. Serotyping and biotyping methods have been commonly used for identifying bacterial isolates and for epideomiological purposes in the past and currently in some countries [155]. These phenotypic methods, however, cannot provide as much discriminatory power as genotyping methods. The MLST, PFGE and AFLP have been found to have greater discriminatory powers when compared with techniques like ribotyping and flagellin typing. It is not yet possible to identify a perfect typing method for all non-pathogenic and pathogenic *Campylobacter* species. However, currently available techniques including NGS, when used in concert, would fulfill the requirements for epidemiological and source attribution purposes. The development of a validated and practical typing method or methods could make routine subtyping of *Campylobacter* species feasible.

4. Discussion and conclusion

Currently, the genus *Campylobacter* contains 39 validly published species, 11 subspecies and 4 biovars, according to the list of prokaryotic names with standing in nomenclature. These diversified species of the genus *Campylobacter* transit through various animal and environmental compartments to humans and animals, which emphasize the need to adopt an integrated One-Health approach in *Campylobacter* epidemiology, risk assessment and prevention. Thermotolerant campylobacters, such as *C. jejuni*, *C. coli* and *C. lari*, are the most implicated species in *Campylobacter* infections in humans. However, there are many other emerging and unusual species, which cannot be detected using the currently available culture methods. Problems with recovering different strains within the same species or specific relevant strains using the same media formulation are often encountered because of the multiple resistance of campylobacters to antibiotics. Many studies highlight the importance of method sensitivity and the need to collect multiple isolates from both clinical samples and potential sources of infection to support epidemiological investigations. All the above-mentioned testing limitations may have contributed to missed or inadequate source attribution and a limited understanding of the epidemiology of *Campylobacter* gastroenteritis. Therefore, the development of the new strategies, employing comprehensive procedures to isolate and detect all the strains and species of campylobacters would be ideal. In order to prevent and control the transmission of foodborne campylobacters to humans, the baseline studies and risk assessments at larger scales with more systemic and collaborative approaches have been strengthened in recent years. The measures to reduce the burden of campylobacters in poultry products have been implemented in several countries with success, including the regulatory requirements to meet the limit of *Campylobacter* load on the carcasses of young chicken and turkeys in the US and Europe. To overcome difficulties in preserving most fresh foods with short shelf-life, the food industry urgently demands novel rapid tests at reasonable cost that employ improved culture and culture-independent methods able to accurately detect low numbers of viable *Campylobacter* cells. These innovative methods will reduce significant economic loss and the health risks to public.

Acknowledgements

The authors thank Beverley Phipps-Todd for providing laboratory support throughout many years in developing procedures for the isolation and detection of campylobacters, and Sohail Naushad, Ruimin Gao and Isaac Firth, for providing critical review of this manuscript.

Conflict of interest

The authors declare no conflict of interest.

Author details

Honsheng Huang[1*] and Manuel Mariano Garcia[2,3*]

1 Ottawa Laboratory – Fallowfield, Canadian Food Inspection Agency, Ottawa, Ontario, Canada

2 Microbiology and Food Safety, Ottawa, Canada

3 National Academy of Science and Technology, Manila, Philippines

*Address all correspondence to: hongsheng.huang@inspection.gc.ca; mannygarcia1967@gmail.com

References

[1] Butzler JP. Campylobacter, from obscurity to celebrity. Clinical Microbiology and Infection. 2004;**10** (10):868-876

[2] Kaakoush NO, Castano-Rodriguez N, Mitchell HM, Man SM. Global epidemiology of Campylobacter infection. Clinical Microbiology Reviews. 2015;**28**(3):687-720

[3] Platts-Mills JA, Kosek M. Update on the burden of Campylobacter in developing countries. Current Opinion in Infectious Diseases. 2014;**27**(5):444-450

[4] WHO. The Global View of Campylobacterosis. Report of an Expert Consultation. Utrecht, Netherlands: WHO; 2012

[5] Kirk MD, Pires SM, Black RE, Caipo M, Crump JA, Devleesschauwer B, et al. World Health Organization estimates of the global and regional disease burden of 22 foodborne bacterial, protozoal, and viral diseases, 2010: A data synthesis. PLOS Medicine. 2015;**12**(12):e1001921

[6] Vandamme P, Dewhirst F, Paster B, On S. Campylobacteraceae. In: Garrity GM, Brenner DJ, Krieg NR, JT S, editors. In: Bergey's Manual of Systematic Bacteriology, 2. New York, NY: Springer Science; 2005. p. 1147–1160.

[7] Man SM. The clinical importance of emerging Campylobacter species. Nature Reviews Gastroenterology & Hepatology. 2011;**8**(12):669-685

[8] Huang H, Brooks BW, Lowman R, Carrillo CD. Campylobacter species in animal, food, and environmental sources, and relevant testing programs in Canada. Canadian Journal of Microbiology. 2015;**61**(10):701-721

[9] Gaynor EC, Wells DH, MacKichan JK, Falkow S. The Campylobacter jejuni stringent response controls specific stress survival and virulence-associated phenotypes. Molecular Microbiology. 2005;**56**(1):8-27

[10] Tresse O, Alvarez-Ordonez A, Connerton IF. Editorial: About the foodborne pathogen Campylobacter. Frontiers in Microbiology. 2017;**8**:1908

[11] Silva J, Leite D, Fernandes M, Mena C, Gibbs PA, Teixeira P. Campylobacter spp. as a foodborne pathogen: A review. Frontiers in Microbiology. 2011;**2**:200

[12] Lastovica AJ, On SLW, Zhang L. The family Campylobacteraceae. In: Rosenberg E, DeLong EF, Lory S, Stackebrandt E, Thompson F, editors. The Prokaryotes: Deltaproteobacteria and Epsilonproteobacteria. Berlin, Heidelberg: Springer Berlin Heidelberg; 2014. pp. 307-335

[13] Moore JE, Corcoran D, Dooley JS, Fanning S, Lucey B, Matsuda M, et al. Campylobacter. Veterinary Research. 2005;**36**(3):351-382

[14] Skirrow MB. John McFadyean and the centenary of the first isolation of Campylobacter species. Clinical Infectious Diseases. 2006;**43**(9):1213-1217

[15] Smith T. The etiological relation of spirilla (Vibrio fetus) to bovine abortion. Journal of Experimental Medicine. 1919;**30**(4):313-323

[16] Smith T, Taylor MS. Some morphological and biological characters of the spirilla (Vibrio fetus, N. Sp.) associated with disease of the fetal membranes in cattle. Journal of Experimental Medicine. 1919;**30**(4): 299-311

[17] Stenenga T, Terpstra J. Over Vibrio foetus infecties bij het rund en enzootische steriliteit. Tijdschr Diergeneesk. 1949;**74**:293-296

[18] Jones FS, Orcutt M, Little RB. Vibrios (Vibrio Jejuni, N.Sp.) associated with intestinal disorders of cows and calves. Journal of Experimental Medicine. 1931;**53**(6):853-863

[19] Doyle MP. A Vibrio associated with swine dysentery. American Journal of Veterinary Research. 1944; **5**:3-5

[20] Levy AJ. A gastro-enteritis cutbreak probably due to a bovine strain of Vibrio. Yale Journal of Biology and Medicine. 1946;**18**:243-258

[21] Vinzent R, Dumas J, Picard N. Septicémie grave au cours de la grossesse due à un Vibrion. Avortement consécutif. Bulletin de L'Académie Nationale de Médecine. 1947;**131**:90-92

[22] King EO. Human infections with Vibrio fetus and a closely related Vibrio. The Journal of Infectious Diseases. 1957; **101**(2):119-128

[23] Sebald M, Véron M. Teneur en bases de l'ADN et classification des vibrions. The Annales de l'Institut Pasteur (Paris). 1963;**105**:897-910

[24] Véron M, Chatelain R. Taxonomic study of the genus Campylobacter Sebald and Véron and designation of the Neotype strain for the type species, Campylobacter fetus (Smith and Taylor) Sebald and Véron. International Journal of Food Microbiology. 1973; **23**(2):122-134

[25] Dekeyser P, Gossuin-Detrain M, Butzler JP, Sternon J. Acute enteritis due to related Vibrio: First positive stool cultures. The Journal of Infectious Diseases. 1972;**125**(4): 390-392

[26] Butzler JP, Dekeyser P, Lafontaine T. Susceptibility of related vibrios and Vibrio fetus to twelve antibiotics. Antimicrobial Agents and Chemotherapy. 1974;**5**(1):86-89

[27] Skirrow MB. Campylobacter enteritis: A "new" disease. British Medical Journal. 1977;**2**(6078):9-11

[28] Pires SM, Fischer-Walker CL, Lanata CF, Devleesschauwer B, Hall AJ, Kirk MD, et al. Aetiology-specific estimates of the global and regional incidence and mortality of Diarrhoeal diseases commonly transmitted through food. PLoS One. 2015;**10**(12):e0142927

[29] Boukerb AM, Penny C, Serghine J, Walczak C, Cauchie HM, Miller WG, et al. Campylobacter armoricus sp. nov., a novel member of the Campylobacter lari group isolated from surface water and stools from humans with enteric infection. The. International Journal of Systematic and Evolutionary Microbiology. 2019;**69**(12):3969-3979

[30] Bryant E, Shen Z, Mannion A, Patterson M, Buczek J, Fox JG. Campylobacter taeniopygiae sp. nov., Campylobacter aviculae sp. nov., and Campylobacter estrildidarum sp. nov., novel species isolated from laboratory-maintained zebra finches. Avian Diseases. 2020;**64**(4):457-466

[31] Huang H, Carrillo C, Sproston E. Campylobacter. Chapter 27. In: Handbook of Foodborne Diseases. Taylor and Francis CRC Press; 2018. pp. 287-299

[32] Costa D, Iraola G. Pathogenomics of emerging Campylobacter species. Clinical Microbiology Reviews. 2019; **32**(4):e00072-e00018

[33] Moore RJ, Scott PC, Van TTH. Spotlight on avian pathology: Campylobacter hepaticus, the cause of spotty liver disease in layers. Avian Pathology. 2019;**48**(4):285-287

[34] Silva MF, Pereira G, Carneiro C, Hemphill A, Mateus L, Lopes-da-Costa L, et al. Campylobacter portucalensis sp. nov., a new species of Campylobacter

isolated from the preputial mucosa of bulls. PLoS One. 2020;15(1):e0227500

[35] Lee S, Lee J, Ha J, Choi Y, Kim S, Lee H, et al. Clinical relevance of infections with zoonotic and human oral species of Campylobacter. Journal of Microbiology. 2016;**54**(7):459-467

[36] Sahin O, Yaeger M, Wu Z, Zhang Q. Campylobacter-associated diseases in animals. Annual Review of Animal Biosciences. 2017;**5**:21-42

[37] Humphrey T, O'Brien S, Madsen M. Campylobacters as zoonotic pathogens: A food production perspective. The International Journal of Food Microbiology. 2007;**117**(3):237-257

[38] EFSA, ECDC. The European Union summary report on trends and sources of zoonoses, zoonotic agents and food-borne outbreaks in 2014. EFSA Journal. 2015;**13**(12):4329

[39] Mehta R, Singhal P, Singh H, Damle D, Sharma AK. Insight into thermophiles and their wide-spectrum applications. 3. Biotech. 2016;**6**(1):81-81

[40] Christopher FM, Smith GC, Vanderzant C. Effect of temperature and pH on the survival of Campylobacter fetus. Journal of Food Protection. 1982;**45**(3):253-259

[41] Nguyen HT, Corry JE, Miles CA. Heat resistance and mechanism of heat inactivation in thermophilic Campylobacters. Applied and Environmental Microbiology. 2006;**72** (1):908-913

[42] Levin RE. Campylobacter jejuni: A review of its characteristics, pathogenicity, ecology, distribution, subspecies characterization and molecular methods of detection. Food Biotechnology. 2007;**21**(4):271-347

[43] Garcia MM, Lior H, Stewart RB, Ruckerbauer GM, Trudel JR, Skljarevski A. Isolation, characterization, and serotyping of Campylobacter jejuni and Campylobacter coli from slaughter cattle. Applied and Environmental Microbiology. 1985;**49**(3):667-672

[44] Kramer JM, Frost JA, Bolton FJ, Wareing DR. Campylobacter contamination of raw meat and poultry at retail sale: Identification of multiple types and comparison with isolates from human infection. Journal of Food Protection. 2000;**63**(12):1654-1659

[45] Park SF. The physiology of Campylobacter species and its relevance to their role as foodborne pathogens. International Journal of Food Microbiology. 2002;**74**(3):177-188

[46] Murphy C, Carroll C, Jordan KN. Environmental survival mechanisms of the foodborne pathogen Campylobacter jejuni. The Journal of Applied Microbiology. 2006;**100**(4):623-632

[47] Bronowski C, James CE, Winstanley C. Role of environmental survival in transmission of Campylobacter jejuni. FEMS Microbiology Letters. 2014;**356**(1): 8-19

[48] Kim S-H, Chelliah R, Ramakrishnan SR, Perumal AS, Bang W-S, Rubab M, et al. Review on stress tolerance in Campylobacter jejuni. Frontiers in cellular and infection. Microbiology. 2021;**10**

[49] Lee H, Ma R, Grimm MC, Riordan SM, Lan R, Zhong L, et al. Examination of the anaerobic growth of Campylobacter concisus strains. International Journal of Microbiology. 2014;**2014**:476047

[50] Kaakoush NO, Miller WG, De Reuse H, Mendz GL. Oxygen requirement and tolerance of Campylobacter jejuni. Research in Microbiology. 2007;**158**(8–9):644-650

[51] Rodrigues RC, Pocheron AL, Hernould M, Haddad N, Tresse O,

Cappelier JM. Description of Campylobacter jejuni bf, an atypical aero-tolerant strain. Gut Pathogens. 2015;7:30

[52] O'Kane PM, Connerton IF. Characterisation of Aerotolerant forms of a robust chicken colonizing Campylobacter coli. Frontiers in Microbiology. 2017;**8**:513

[53] Forsythe SJ. Introduction to safe food. In: The Microbiology of Safe Food. 2000. pp. 1-9

[54] Piccirillo A, Niero G, Calleros L, Perez R, Naya H, Iraola G. Campylobacter geochelonis sp. nov. isolated from the western Hermann's tortoise (Testudo hermanni hermanni). International Journal of Food Microbiology. 2016;**66**(9):3468-3476

[55] Debruyne L, Gevers D, Vandamme P. Deltaproteobacteria and Epsilonproteobacteria. In: Rosenberg E, DeLong EF, Lory S, Stackebrandt E, Thompson F, editors. The Prokaryotes: Deltaproteobacteria and Epsilonproteobacteria. Berlin, Heidelberg: Springer Berlin Heidelberg; 2014. pp. 307-335

[56] Penner JL. The genus Campylobacter: A decade of progress. Clinical Microbiology Reviews. 1988;**1** (2):157-172

[57] Hazeleger WC, Wouters JA, Rombouts FM, Abee T. Physiological activity of Campylobacter jejuni far below the minimal growth temperature. Applied and Environmental Microbiology. 1998;**64**(10):3917-3922

[58] De Cesare A, Sheldon BW, Smith KS, Jaykus LA. Survival and persistence of Campylobacter and Salmonella species under various organic loads on food contact surfaces. Journal of Food Protection. 2003;**66**(9):1587-1594

[59] Stern NJ, Kotula AW. Survival of Campylobacter jejuni inoculated into ground beef. Applied and Environmental Microbiology. 1982; **44**(5):1150-1153

[60] Lee A, Smith SC, Coloe PJ. Survival and growth of Campylobacter jejuni after artificial inoculation onto chicken skin as a function of temperature and packaging conditions. Journal of Food Protection. 1998;**61**(12): 1609-1614

[61] Corry JE, Post DE, Colin P, Laisney MJ. Culture media for the isolation of Campylobacters. The International Journal of Food Microbiology. 1995; **26**(1):43-76

[62] Thomas GD. Pilot study for the development of a new Campylobacter selective medium at 37 degrees C using aztreonam. The Journal of Clinical Pathology. 2005;**58**(4):413-416

[63] Bourke B, Chan VL, Sherman P. Campylobacter upsaliensis: Waiting in the wings. Clinical Microbiology Reviews. 1998;**11**(3):440-449

[64] Lastovica AJ, Le Roux E, Penner JL. "Campylobacter upsaliensis" isolated from blood cultures of pediatric patients. The Journal of Clinical Microbiology. 1989;**27**(4):657-659

[65] Lastovica AJ, le Roux E. Efficient isolation of Campylobacteria from stools. The Journal of Clinical Microbiology. 2000;**38**(7):2798-2799

[66] Hofreuter D. Defining the metabolic requirements for the growth and colonization capacity of Campylobacter jejuni. Frontiers in Cellular and Infection Microbiology. 2014;**4**:137

[67] Parkhill J, Wren BW, Mungall K, Ketley JM, Churcher C, Basham D, et al. The genome sequence of the food-borne pathogen Campylobacter jejuni reveals hypervariable sequences. Nature. 2000; **403**(6770):665-668

[68] Velayudhan J, Kelly DJ. Analysis of gluconeogenic and anaplerotic enzymes in Campylobacter jejuni: An essential role for phosphoenolpyruvate carboxykinase. Microbiology (Reading). 2002;**148**(Pt 3):685-694

[69] Gundogdu O, Bentley SD, Holden MT, Parkhill J, Dorrell N, Wren BW. Re-annotation and re-analysis of the Campylobacter jejuni NCTC11168 genome sequence. BMC Genomics. 2007;**8**(1):162

[70] Bochner BR. Global phenotypic characterization of bacteria. FEMS microbiology Reviews. 2009;**33**(1): 191-205

[71] Line JE, Hiett KL, Guard-Bouldin J, Seal BS. Differential carbon source utilization by Campylobacter jejuni 11168 in response to growth temperature variation. The Journal of Microbiological Methods. 2010;**80**(2): 198-202

[72] Gripp E, Hlahla D, Didelot X, Kops F, Maurischat S, Tedin K, et al. Closely related Campylobacter jejuni strains from different sources reveal a generalist rather than a specialist lifestyle. BMC Genomics. 2011;**12**:584

[73] Muraoka WT, Zhang Q. Phenotypic and genotypic evidence for L-fucose utilization by Campylobacter jejuni. Journal of Bacteriology. 2011;**193**(5): 1065-1075

[74] El Baaboua A, El Maadoudi M, Bouyahya A, Kounnoun A, Bougtaib H, Omar B, et al. A review of current knowledge and gaps about Campylobacter methods from culture to characterization. Journal of microbiology, biotechnology and food sciences. 2021:e4154

[75] Day CJ, Hartley-Tassell LE, Shewell LK, King RM, Tram G, Day SK, et al. Variation of chemosensory receptor content of Campylobacter jejuni strains and modulation of receptor gene expression under different in vivo and in vitro growth conditions. BMC Microbiology. 2012;**12**(1):128

[76] Rahman H, King RM, Shewell LK, Semchenko EA, Hartley-Tassell LE, Wilson JC, et al. Characterisation of a multi-ligand binding chemoreceptor CcmL (Tlp3) of Campylobacter jejuni. PLOS Pathogens. 2014;**10**(1):e1003822

[77] Gharst G, Oyarzabal OA, Hussain SK. Review of current methodologies to isolate and identify Campylobacter spp. from foods. The Journal of Microbiological Methods. 2013;**95**(1): 84-92

[78] Oakley BB, Morales CA, Line JE, Seal BS, Hiett KL. Application of high-throughput sequencing to measure the performance of commonly used selective cultivation methods for the foodborne pathogen Campylobacter. FEMS Microbiology Ecology. 2012; **79**(2):327-336

[79] Whitehouse CA, Young S, Li C, Hsu CH, Martin G, Zhao S. Use of whole-genome sequencing for Campylobacter surveillance from NARMS retail poultry in the United States in 2015. Food Microbiology. 2018;**73**:122-128

[80] Vidal AB, Colles FM, Rodgers JD, McCarthy ND, Davies RH, Maiden MCJ, et al. Genetic diversity of Campylobacter jejuni and Campylobacter coli isolates from conventional broiler flocks and the impacts of sampling strategy and laboratory method. Applied and Environmental Microbiology. 2016; **82**(8):2347-2355

[81] Joensen KG, Kuhn KG, Muller L, Bjorkman JT, Torpdahl M, Engberg J, et al. Whole-genome sequencing of Campylobacter jejuni isolated from Danish routine human stool samples reveals surprising degree of clustering. Clinical Microbiology and Infection. 2018;**24**(2):201 e205-201 e208

[82] Tong S, Ma L, Ronholm J, Hsiao W, Lu X. Whole genome sequencing of Campylobacter in Agri-food surveillance. Current Opinion in Food Science. 2021;**39**:130-139

[83] On SLW. Isolation, identification and subtyping of Campylobacter: Where to from here? The Journal of Microbiological Methods. 2013;**95**(1): 3-7

[84] Smith MV 2nd, Muldoon PJ. Campylobacter fetus subspecies jejuni (Vibrio fetus) from commercially processed poultry. Journal of Applied Microbiology. 1974;**27**(5):995-996

[85] Corry JEL, Atabay HI. Chapter 19 culture Media for the Isolation of Campylobacters, Helicobacters and Arcobacters. In: Handbook of Culture Media for Food and Water Microbiology. Vol. 3. The Royal Society of Chemistry; 2012. pp. 403-450

[86] Beuchat LR. Efficacy of media and methods for detecting and enumerating Campylobacter jejuni in refrigerated chicken meat. Applied and Environmental Microbiology. 1985;**50** (4):934-939

[87] Donnison A. Isolation of Thermotolerant Campylobacter – Review & Methods for New Zealand laboratories, client report/ministry of. Health. 2003

[88] Lovett J, Francis DW, Hunt JM. Isolation of Campylobacter jejuni from raw milk. Applied and Environmental Microbiology. 1983;**46**(2):459-462

[89] USDA-FSIS-2021. Isolation, identification, and enumeration of Campylobacter jejuni/coli/lari from poultry rinse and sponge samples. In: Laboratory Guidebook, MLG 41.00. 2021. Available from: https://www.fsis.usda.gov/sites/default/files/media_file/2021-08/MLG-41.06

[90] Oyarzabal OA, Backert S, Nagaraj M, Miller RS, Hussain SK, Oyarzabal EA. Efficacy of supplemented buffered peptone water for the isolation of Campylobacter jejuni and C. coli from broiler retail products. The. Journal of Microbiological Methods. 2007;**69**(1): 129-136

[91] FDA-BAM. Bacteriological analytical manual (BAM). Chapter 7. In: Hunt JM, Abeyta C, Tran T, editors. Campylobacter. 8th edition (revision A) ed. FDA-BAM; 23 pages; 2000. Available from: https://www.fda.gov/food/laboratory-methods-food/bamchapter-7-CampylobacterContent current as of: 08/03/2021

[92] OIE-2017. World Organization for Animal Health. OIE terrestrial manual 2017. Chapter 2.9.3. In: Infection with Campylobacter Jejuni and C. coli. World Organization for Animal Health. 2017. Available from: https://www.oie.int/fileadmin/Home/eng/Health_standards/tahm/2.09.03_CAMPYLO

[93] Stern NJ, Wojton B, Kwiatek K. A differential-selective medium and dry ice-generated atmosphere for recovery of Campylobacter jejuni. Journal of Food Protection. 1992;**55**(7):514-517

[94] Line J, Svetoch E, Eruslanov B, Perelygin V, Mitsevich E, Mitsevich I, et al. Isolation and purification of enterocin E-760 with broad antimicrobial activity against gram-positive and gram-negative bacteria. Antimicrobial Agents and Chemotherapy. 2008;**52**(3):1094-1100

[95] Public-Health-England-2018. Detection and Enumeration of Campylobacter Species National Infection Service Food Water and Environmental Microbiology Standard Method, Document number FNES15 (F21) Version No 4 Effective Date: 27.04.2018. Available from: https://assets.publishing.service.gov.uk/government/uploads/system/uploads/attachme

nt_data/file/866763/National_SOP_FNES15_F21_Detection_and_Enumeration_of_Campylobacter_Species

[96] Seliwiorstow T, Bare J, Verhaegen B, Uyttendaele M, de Zutter L. Evaluation of a new chromogenic medium for direct enumeration of Campylobacter in poultry meat samples. Journal of Food Protection. 2014;**77**(12): 2111-2114

[97] Rosenquist H, Bengtsson A, Hansen TB. A collaborative study on a Nordic standard protocol for detection and enumeration of thermotolerant Campylobacter in food (NMKL 119, 3. Ed., 2007). International Journal of Food Microbiology. 2007;**118**(2):201-213

[98] Ricke SC, Feye KM, Chaney WE, Shi Z, Pavlidis H, Yang Y. Developments in rapid detection methods for the detection of foodborne Campylobacter in the United States. Frontiers in Microbiology. 2019;**9**:3280

[99] ISO-10272-1:2017. International Organization for Standardization. ISO 10272-1:2017. Microbiology of the Food Chain — Horizontal Method for Detection and Enumeration of Campylobacter spp. — Part 1: Detection Method. International Organization for Standardization; 2017

[100] ISO-10272-1-2:2017. Microbiology of the Food Chain—Horizontal Method for Detection and Enumeration of Campylobacter spp. — Part 2: Colony-Count Technique. ISO-10272-1-2:2017; 2017

[101] Health-Canada-2014. Laboratory Procedure MFLP - 46, March 2014, Isolation of Thermophilic Campylobacter in Foods, Compendium of Analytic Methods. Laboratory Procedure MFLP – 46. Health-Canada-2014; 2014

[102] FSANZ. Food Standards Australia New Zealand (FSANZ). Compendium of Microbiological Criteria for Food (Revised Jan 2018). FSANZ; 2016

[103] Engberg J, On SL, Harrington CS, Gerner-Smidt P. Prevalence of Campylobacter, Arcobacter, Helicobacter, and Sutterella spp. in human fecal samples as estimated by a reevaluation of isolation methods for Campylobacters. Journal of Clinical Microbiology. 2000;**38**(1):286-291

[104] Speegle L, Miller ME, Backert S, Oyarzabal OA. Use of cellulose filters to isolate Campylobacter spp. from naturally contaminated retail broiler meat. Journal of Food Protection. 2009; **72**(12):2592-2596

[105] Valdivieso-Garcia A, Harris K, Riche E, Campbell S, Jarvie A, Popa M, et al. Novel Campylobacter isolation method using hydrophobic grid membrane filter and semisolid medium. Journal of Food Protection. 2007;**70**(2): 355-362

[106] Jokinen CC, Koot JM, Carrillo CD, Gannon VP, Jardine CM, Mutschall SK, et al. An enhanced technique combining pre-enrichment and passive filtration increases the isolation efficiency of Campylobacter jejuni and Campylobacter coli from water and animal fecal samples. The Journal of Microbiological Methods. 2012;**91**(3):506-513

[107] Chon JW, Jung JY, Ahn Y, Bae D, Khan S, Seo KH, et al. Detection of Campylobacter jejuni from fresh produce: Comparison of culture- and PCR-based techniques, and metagenomic approach for analyses of the microbiome before and after enrichment. Journal of Food Protection. 2021;**84**(10):1704-1712

[108] Bolton FJ, Hutchinson DN, Coates D. Blood-free selective medium for isolation of Campylobacter jejuni from feces. Journal of Clinical Microbiology. 1984;**19**(2):169-171

[109] Lynch OA, Cagney C, McDowell DA, Duffy G. A method for the growth and recovery of 17 species of Campylobacter and its subsequent application to inoculated beef. The Journal of Microbiological Methods. 2010;**83**(1):1-7

[110] Kim J, Oh E, Banting GS, Braithwaite S, Chui L, Ashbolt NJ, et al. An improved culture method for selective isolation of Campylobacter jejuni from wastewater. Frontiers in Microbiology. 2016;**7**:1345

[111] Kim J, Shin H, Park H, Jung H, Kim J, Cho S, et al. Microbiota analysis for the optimization of Campylobacter isolation from chicken carcasses using selective media. Frontiers in Microbiology. 2019;**10**:1381

[112] Huang H, Phipps-Todd B. Improvement of capture efficacy of immunomagnetic beads for Campylobacter jejuni using reagents that alter its motility. Canadian Journal of Microbiology. 2013;**59**(7): 511-514

[113] Hunt J, Abeyta C, Tran T. Isolation of Campylobacter species from food and water. Bacteriological Analytical Manual (BAM). 2001:7.01-07.24

[114] Reilly SS, Gilliland SE. Improved culturing techniques for Campylobacter. Food Science. 2003;**68**(9):2752-2757

[115] Oyarzabal OA, Macklin KS, Barbaree JM, Miller RS. Evaluation of agar plates for direct enumeration of Campylobacter spp. from poultry carcass rinses. Applied and Environmental Microbiology. 2005;**71** (6):3351-3354

[116] Szalanski AL, Owens CB, McKay T, Steelman CD. Detection of Campylobacter and Escherichia coli O157:H7 from filth flies by polymerase chain reaction. Medical and Veterinary Entomology. 2004;**18**(3):241-246

[117] Davis L, DiRita V. Growth and laboratory maintenance of Campylobacter jejuni. Current Protocols in Microbiology. 2008;Chapter 8:Unit 8A:1 1-8A 1 7

[118] Buswell CM, Nicholl HS, Walker JT. Use of continuous culture bioreactors for the study of pathogens such as Campylobacter jejuni and Escherichia coli O157 in biofilms. Methods in Enzymology. 2001;**337**: 70-78

[119] Clarke AK, Ajlouni S. Recommended practices to eliminate Campylobacter from live birds and chicken meat in Japan. Food Safety (Tokyo). 2021;**9**(3):57-74

[120] United-States-Department-of-Agriculture-2010. New Performance Standards for Salmonella and Campylobacter in Young Chicken and Turkey Slaughter Establishments; New Compliance Guides. Federal Register 2010. Vol. 75. United-States-Department-of-Agriculture, No. 93/May 14, 2010/Notices. Available from: http://www.fsis.usda.gov/OPPDE/rdad/FRPubs/2009-0034; 2010

[121] European-Commission-2017. COMMISSION REGULATION (EU) 2017/1495, amending Regulation (EC), No 2073/2005 COMMISSION REGULATION (EU) 2017/1495, of 23 august 2017, as regards Campylobacter in broiler car cases EU 2017/1495, Commission Regulation No. 2073/2005 of 15 November 2005 on microbiological criteria for foodstuffs (OJ L 338, 22.12.2005, p. 1). Official Journal of the European Union L 218/1. 2017. Available from: https://eur-lex.europa.eu/ell/reg/2017/1495/oj

[122] Scallan E, Griffin PM, Angulo FJ, Tauxe RV, Hoekstra RM. Foodborne illness acquired in the United States–unspecified agents. Emerging Infectious Diseases. 2011;**17**(1):16-22

[123] Scallan E, Hoekstra RM, Angulo FJ, Tauxe RV, Widdowson MA, Roy SL, et al. Foodborne illness acquired in the United States–major pathogens. Emerging Infectious Diseases. 2011;**17**(1):7-15

[124] Bélanger P, Tanguay F, Hamel M, Phypers M. Outbreak. Report an Overview of Foodborne Outbreaks in Canada Reported through Outbreak Summaries: 2008–2014. Vol. 254. Canada Communicable Disease Report; 2015. pp. 41-11. DOI: 10.14745/ccdr.v41i11a01

[125] Inglis GD, Boras VF, Houde A. Enteric Campylobacteria and RNA viruses associated with healthy and diarrheic humans in the Chinook health region of southwestern Alberta, Canada. Journal of Clinical Microbiology. 2011;**49**(1):209-219

[126] Gilpin B, Robson B, Lin S, Scholes P, On S. Pulsed-field gel electrophoresis analysis of more than one clinical isolate of Campylobacter spp. from each of 49 patients in New Zealand. Journal of Clinical Microbiology. 2012;**50**(2):457-459

[127] Buss JE, Cresse M, Doyle S, Buchan BW, Craft DW, Young S. Campylobacter culture fails to correctly detect Campylobacter in 30% of positive patient stool specimens compared to non-cultural methods. European Journal of Clinical Microbiology & Infectious Diseases. 2019;**38**(6):1087-1093

[128] Kulkarni SP, Lever S, Logan JM, Lawson AJ, Stanley J, Shafi MS. Detection of Campylobacter species: A comparison of culture and polymerase chain reaction based methods. The Journal of Clinical Pathology. 2002;**55**(10):749-753

[129] Bolton FJ, Sails AD, Fox AJ, Wareing DR, Greenway DL. Detection of Campylobacter jejuni and Campylobacter coli in foods by enrichment culture and polymerase chain reaction enzyme-linked immunosorbent assay. Journal of Food Protection. 2002;**65**(5):760-767

[130] Oyarzabal OA, Williams A, Zhou P, Samadpour M. Improved protocol for isolation of Campylobacter spp. from retail broiler meat and use of pulsed field gel electrophoresis for the typing of isolates. The Journal of Microbiological Methods. 2013;**95**(1):76-83

[131] Petersen L, Nielsen EM, On SL. Serotype and genotype diversity and hatchery transmission of Campylobacter jejuni in commercial poultry flocks. Veterinary Microbiology. 2001;**82**(2):141-154

[132] Nadeau E, Messier S, Quessy S. Prevalence and comparison of genetic profiles of Campylobacter strains isolated from poultry and sporadic cases of Campylobacteriosis in humans. Journal of Food Protection. 2002;**65**(1):73-78

[133] Newell DG, Fearnley C. Sources of Campylobacter colonization in broiler chickens. Applied and Environmental Microbiology. 2003;**69**(8):4343-4351

[134] Ellerbroek LI, Lienau JA, Klein G. Campylobacter spp. in broiler flocks at farm level and the potential for cross-contamination during slaughter. Zoonoses and Public Health. 2010;**57**(7–8):e81-e88

[135] Gruntar I, Biasizzo M, Kušar D, Pate M, Ocepek M. Campylobacter jejuni contamination of broiler carcasses: Population dynamics and genetic profiles at slaughterhouse level. Food Microbiology. 2015;**50**:97-101

[136] Williams LK, Sait LC, Cogan TA, Jorgensen F, Grogono-Thomas R, Humphrey TJ. Enrichment culture can bias the isolation of Campylobacter subtypes. Epidemiology and Infection. 2012;**140**(7):1227-1235

[137] Huang H, Phipps-Todd B, McMahon T, Elmgren CL, Lutze-Wallace C, Todd ZA, et al. Development of a monoclonal antibody-based colony blot immunoassay for detection of thermotolerant Campylobacter species. The Journal of Microbiological Methods. 2016;**130**:76-82

[138] Ugarte-Ruiz M, Wassenaar TM, Gomez-Barrero S, Porrero MC, Navarro-Gonzalez N, Dominguez L. The effect of different isolation protocols on detection and molecular characterization of Campylobacter from poultry. Letters in Applied Microbiology. 2013;**57**(5):427-435

[139] Black RE, Levine MM, Clements ML, Hughes TP, Blaser MJ. Experimental Campylobacter jejuni infection in humans. The Journal of Infectious Diseases. 1988;**157**(3):472-479

[140] Robinson DA. Infective dose of Campylobacter jejuni in milk. British Medical Journal (clinical research edition). 1981;**282**(6276):1584

[141] Huang H, Raymond P, Grenier C, Fahey J. Chapter 7 - development and improvement of a colony blot immunoassay for the detection of thermotolerant Campylobacter species. Methods in Microbiology. 2020; **47**:209-244

[142] Bolton FJ, Holt AV, Hutchinson DN. Campylobacter biotyping scheme of epidemiological value. The Journal of Clinical Pathology. 1984;**37**(6):677-681

[143] Skirrow MB, Benjamin J. Differentiation of enteropathogenic Campylobacter. The Journal of Clinical Pathology. 1980;**33**(11):1122

[144] Lior H, Woodward DL, Edgar JA, Laroche LJ, Gill P. Serotyping of Campylobacter jejuni by slide agglutination based on heat-labile antigenic factors. The Journal of Clinical Microbiology. 1982;**15**(5):761-768

[145] Penner JL, Hennessy JN. Passive hemagglutination technique for serotyping Campylobacter fetus subsp. jejuni on the basis of soluble heat-stable antigens. The Journal of Clinical Microbiology. 1980;**12**(6):732-737

[146] Tenover FC, Williams S, Gordon KP, Harris N, Nolan C, Plorde JJ. Utility of plasmid fingerprinting for epidemiological studies of Campylobacter jejuni infections. The Journal of Infectious Diseases. 1984; **149**(2):279

[147] Bradbury WC, Marko MA, Hennessy JN, Penner JL. Occurrence of plasmid DNA in serologically defined strains of Campylobacter jejuni and Campylobacter coli. Infection and Immunity. 1983;**40**(2):460-463

[148] Grajewski BA, Kusek JW, Gelfand HM. Development of a bacteriophage typing system for Campylobacter jejuni and Campylobacter coli. Journal of Clinical Microbiology. 1985;**22**(1):13-18

[149] Eberle KN, Kiess AS. Phenotypic and genotypic methods for typing Campylobacter jejuni and Campylobacter coli in poultry. Poultry Science. 2012;**91**(1):255-264

[150] Taboada EN, Clark CG, Sproston EL, Carrillo CD. Current methods for molecular typing of Campylobacter species. The Journal of Microbiological Methods. 2013;**95**(1):24-31

[151] Wassenaar TM, Newell DG. Genotyping of Campylobacter spp. Applied and Environmental Microbiology. 2000;**66**(1):1-9

[152] Nennig M, Llarena A-K, Herold M, Mossong J, Penny C, Losch S, et al. Investigating major recurring Campylobacter jejuni lineages in Luxembourg using four Core or whole genome sequencing typing schemes. Frontiers in Cellular and Infection Microbiology. 2021;**10**(818):608020

[153] Quainoo S, Coolen JPM, van Hijum S, Huynen MA, Melchers WJG, van Schaik W, et al. Whole-genome sequencing of bacterial pathogens: The future of nosocomial outbreak analysis. Clinical Microbiology Reviews. 2017; **30**(4):1015-1063

[154] Arning N, Sheppard SK, Bayliss S, Clifton DA, Wilson DJ. Machine learning to predict the source of Campylobacteriosis using whole genome data. PLoS genetics. 2021;**17** (10):e1009436-e1009436

[155] Clarke TN, Schilling MA, Melendez LA, Isidean SD, Porter CK, Poly FM. A systematic review and meta-analysis of Penner serotype prevalence of Campylobacter jejuni in low- and middle-income countries. PLoS One. 2021;**16**(5):e0251039

Chapter 5

Conventional and Molecular Detection Methods of the Opportunistic Bacterial Pathogen *Campylobacter concisus*

Mohsina Huq and Taghrid Istivan

Abstract

Campylobacter concisus is an emerging pathogen that causes gastroenteritis and is a suspected cause of inflammatory bowel diseases. Its importance is enhanced by the chronic sequela that results from acute infection. This bacterium has been under-diagnosed in intestinal infectious diseases, and its clinical importance has not been determined yet. In order to establish the implication of this emerging bacterial species in human gastroenteritis and other infections, different approaches and procedure have been performed, where molecular typing methods have played a central role. The chapter provides a comprehensive past and recent updates on the detection of *C. concisus* by biochemical and molecular methods.

Keywords: *Campylobacter concisus*, hydrogen-requiring microaerophilic, opportunistic pathogen, PCR, PCR-DGGE, MALDI-TOF

1. Introduction

Campylobacter concisus is a fastidious, microaerophilic and hydrogen-requiring mesophile. It is a Gram-negative curved rod bacterium that is normally found in the human oral cavity and is actively motile with a single polar flagellum [1] with a cell size of (0.5–1) × (2–6) μm [2]. This small, non-pigmenting, asaccharolytic bacterium [3] usually grows slowly and requires enriched media. Such characteristics could be linked to its small genome that has a low G + C content (34% -38%) [3]. Unlike other *Campylobacter* spp., *C. concisus* does not have any known primary animal reservoir yet. Probably the human gastrointestinal tract is its only habitat where potential infections may spread via the inter-personal route. However it is worth to note that *C. concisus* was reported to be isolated from slaughtered porcine samples but not from live animals [4], while its DNA has been detected in few animal sources, such as the saliva of domestic pets using PCR-DGGE method [5], and in diarrheic faecal samples from domestic dogs by quantitative PCR [6].

Historically, in 1981 Tanner et al. [3] first recognised *C. concisus* as a member of the microflora of the oral cavity. Oral cavity parts are lips, buccal mucosa, teeth, gums, tongue, the floor of the mouth below the tongue, and the hard bony roof and soft palate. Twenty years later, in addition to its status as a coloniser of the oral cavity in humans, *C. concisus* (with other *Campylobacter* spp.) was considered as an opportunistic pathogen under certain medical conditions by Macuch, Tanner [7]. Since the recognition, there have been many approaches to detect and isolate this bacterium, but it remains unclear whether *C. concisus* is an opportunistic pathogen of inflamed tissues, an oral pathogen, or is simply a commensal of the oral cavity. Here we focus on past and current isolation and detection techniques used to detect this bacterium in clinical samples and other sites of the human's gastrointestinal tract.

2. *C. concisus* in the oral cavity

Several studies examined the composition of the subgingival microbiota of children and found that the detection rate of *C. concisus* in permanent teeth is significantly higher than that of indeciduous teeth ($p < 0.001$) [8, 9]. The prevalence of *C. concisus* in the human oral cavity was detected by a PCR targeting the 16S rRNA gene and it was found in 100% (11/11) of saliva samples collected from healthy individuals [5]. Similar outcomes have been reported by Dewhirst et al. [10], when both cultivation and molecular methods were used to identify the human's oral microbiota. Another study conducted by Zhang et al. [11] also suggested that *C. concisus* was commonly present in the human oral cavity. The study reported the isolation of *C. concisus* from saliva of healthy controls, where 75% (44/59) were culture positive and 97% (57/59) were PCR positive. It is worth to mention here that our research team has successfully isolated this bacterium from 100% healthy human adults, using conventional culture and molecular techniques (unpublished data). These collective data indicate that the human oral cavity is the primary colonisation site of this bacterium.

The association of *C. concisus* with human periodontal diseases is also well reported [3, 9, 12], for example this bacterium was found attached to the teeth in higher numbers than other sites of the oral cavity in patients with periodontitis [13]. Immune responses against *C. concisus* in persons with periodontal diseases were also investigated, with higher antibody levels detected in periodontally diseased subjects compared to healthy controls [14]. Yet in other studies the bacterium was reported to be associated with gingivitis, periodontal sites in addition to healthy sites [15]. *C. concisus* was also reported to be detected in bleeding sites more than non-bleeding sites in periodontitis [8, 9, 16], and in enlarged lesions of gingivitis [17]. The same group reported in a later study that it was more associated with periodontitis in smokers than non-smokers participants [18]. The association of *C. concisus* with periodontitis was also supported by significantly higher isolation rates when gingival crevicular fluids (GCF) of patients were positive for aspartate aminotransferase (AST) compared to patients with negative result for AST [19]. Later, *C. concisus* was included into one of the six successional complexes that are believed to be involved in periodontal diseases [20].

3. *C. concisus* in acute gastroenteritis and chronic gastrointestinal diseases

The correlation between *C. concisus* and gastroenteritis was first reported in 1989 by Vandamme et al. [21]. *C. concisus* along with other *Campylobacter* spp. such as *C. upsaliensis, C. hyointestinalis,* and *C. fetus* have been reported as causative agents of gastroenteritis, but the bacterium remained unidentified when conventional culture techniques with antibiotics in the culture medium were used [22–24]. The introduction of hydrogen to the microaerophilic incubation conditions significantly improved the isolation rates of *C. concisus* from patients with diarrhoea [25]. In 1995 a study in Sweden reported that 6% of the total cases of children with diarrhoea were found to be related to *C. concisus* [26]. Another study conducted in the same year, in Melbourne, reported that 56% of the "*Campylobacter* like organisms" isolated from children with diarrhoea, were identified to be *C. concisus* by conventional culture techniques [27]. The identity of these isolates was also confirmed by molecular techniques in a following study which concluded that *C. concisus* was associated with diarrhoea particularly in infants between 0–35 months of age [28]. Van Etterijck et al. [29] did not report a significant difference in *C. concisus* isolation rate from children with and without diarrhoea (9% in control and 13.2% in patients). However, other studies reported that this bacterium is associated with gastroenteritis cases in children [30–32]. Furthermore, *C. concisus* DNA has been detected in stool samples of patients with gastroenteritis in several studies [24, 33–36], with Nielsen et al. [37] reporting the incidence of *C. concisus* in patients with gastroenteritis, almost as high as the common *C. jejuni* or *C. coli* in a population-based study in Denmark. A more recent study associated *C. concisus* with travellers' diarrhoea in Nepal using 16S *rRNA* PCR of *Campylobacter* [38].

C. concisus was linked to inflammatory bowel diseases (IBD) since 2009 [11] when it was first isolated from stool samples of IBD patients [39]. Furthermore, Zhang et al. [11] found a significantly higher prevalence of *C. concisus* in children with CD than in controls ($p < 0.001$). In 2010, *C. concisus* DNA was detected in faecal samples of CD patients, in a significantly higher ratio (65%) than that of healthy and non-IBD controls (33%) [40]. Hence, in 2011, *C. concisus* was considered to be associated with UC cases [41]. Furthermore, the prevalence of *C. concisus* DNA was significantly higher in biopsy specimens ($p = 0.0019$) of adult UC patients (33.3%) as compared with controls (10.8%), which was supported by another study in the same year [42]. Ismail et al. [43] compared enteric and oral *C. concisus* isolates from eight patients with IBD (four UC and four CD) and six controls by multi-locus sequence typing (MLST), invasion assays, protein analysis, and scanning electron microscopy. Interestingly, the MLST results showed that the majority (87.5%) of *C. concisus* isolates from IBD patients were in one cluster compared to those from the control group (28.6%) ($p < 0.05$). This study provided the first evidence that patients with IBD are colonised with specific oral *C. concisus* strains and these strains may undergo natural recombination. Exotoxin 9, a putative virulence factor which may be associated with increased survival in the cell [44], and the zonula tight junction occludens toxin (Zot) [45] have been associated with virulence properties of *C. concisus* isolates from IBD cases.

4. Identification

4.1 Laboratory diagnosis, isolation, and detection of *C. concisus* in clinical specimens

Since *C. concisus* is usually present with other commensal microorganisms, the filtration culture techniques and/or molecular identification methods are more reliable than standard culture methods.

4.2 Culture and incubation conditions

C. concisus is routinely cultured on Columbia agar base or blood agar base supplemented with 5–6% defibrinated horse blood (HBA) in a special gas mixture containing 7% H_2, 7% CO_2, 5–7% O_2 and ~ 79% N_2 in an anaerobic jar incubated for 48–72 hours at 37 °C [28, 46, 47]. The microaerophilic growth conditions can also be generated by evacuating an anaerobic jar to −7 bar and then gassing with a mixture of 10% H_2, 10% CO_2 and ~ 80% N_2 [46, 47]. *C. concisus* appears on HBA as colonies measuring 1–2 mm in diameter, round, entire, semi translucent and grey in colour [28].

4.3 Identification by cultural and biochemical properties

As *C. concisus* is a fastidious slow growing bacterium that is biochemically inert or inactive, it has been under-reported due to difficulties in isolation and improper identification. Sensitivity to cephalothin and nalidixic acid, growth temperature and colony colour have been used to identify *C. concisus* [48]. Arylsulfatase activity test is another important test, used to differentiate it from *C. mucosalis* and *C. upsaliensis* [49]. *C. concisus* was misidentified as *C. mucosalis* when initially isolated from samples other than the oral cavity [21]. Now, *C. concisus* is reported more often from patients with diarrhoea and other sites because of improvement of the culture system and the use of the stool filtration technique named the "Cape Town Protocol" [50]. In Cape Town, South Africa, the identification rate was reported to be increased by 31% when this technique was used with incubation in a hydrogen-enriched environment [31, 51]. However, as mentioned earlier, recently Nielsen et al. [52] demonstrated the polycarbonate filter is superior to the cellulose acetate filter for detection of *C. concisus*.

The phenotypic characteristics used to identify *C. concisus* in several studies are listed in **Table 1**.

Test	[3]	[2]	[53]	[54]
Active motility	+	+	+	+
Oxidase	+	+ (60–93%)	+	V
Catalase	—	—	—	—
Urease	—	—	—	NA
Hippurate hydrolysis	NA	—	—	—
Benzidine reaction	+			
Indoxyl acetate hydrolysis	NA	—	—	—
Nitrate reduction	+	+ (14–50%)	+	(−)
Selenite reduction		+ (14–50%)		NA
H_2S/TSI	+	—[b]	−/+	NA
Benzyl viologen reduction	+		NA	NA
Neutral red reduction	+		NA	NA
Growth at 25 °C	NA	—	—	
Growth at 42 °C	NA	+ (60–93%)	—	(+)
Growth stimulated by formate and fumarate	+		NA	NA
Alpha-hemolysis	NA		NA	—
MacConkey agar	NA		NA	—
Nutrient agar	—		NA	—
Growth on minimal media	—	—	NA	
NaCl (2.0%)	NA	+ (14–50%)	NA	—

Test	[3]	[2]	[53]	[54]
Glycine (1.0%)	NA		NA	—
Safranin (0.02%)	+	+ (14–50%)	NA	v
Sodium deoxycholate (0.1%)	+	+ (14–50%)	NA	—
Nalidixic acid (32 mg/ml)	+	+ (60–93%)	R	+
Cephalothin (32 mg/ml)	NA	NA	S	—
Metronidazole (4 mg/ml)	MIC: 0.5–2	+ (14–50%)	NA	—

[b]*Trace quantities, (−) most strains are negative, (+) most strains are positive, v variable, R Resistant, S Sensitive, NA not available.*

Table 1.
Biochemical characteristics of C. concisus.

4.4 Isolation from clinical samples

There is no standard technique for the isolation of *C. concisus* from faeces, saliva or tissue. However, the most common technique used to isolate this bacterium from faeces is the 'Cape Town protocol', which involves filtration of samples onto enriched media such as HBA containing antibiotics or onto antibiotic free HBA [31, 32]. Initially the faecal sample is suspended in liquid medium or phosphate buffered saline (PBS) at 1:2 to 1:10, then, 4–5 drops are placed on a cellulose acetate filter (pores size 0.65 μm) positioned on HBA. The soaked filter should be kept on the medium for approximately 10 min to allow the small sized bacterial cells to pass through its pores. Once the filter is discarded, a streak dilution of the primary inoculum is performed then the plate is incubated for 3–5 days in the gas mixture conditions as previously explained [46].

To isolate *C. concisus* from tissue samples such as intestinal biopsies, the homogenised sample is spread on HBA plates containing 10 mg/ml of each trimethoprim and vancomycin prior to incubation under the suitable growth conditions [11, 42]. Alternatively, a two-step enrichment-filtration method can be used [55] as follows: Step 1, the biopsy is enriched by initial incubation for 48 h in microaerophilic conditions in a tube containing 3 ml of Ham's F-12 medium with foetal bovine serum (5% FBS) and 10 μg/ml of vancomycin; step 2, filtration of 200 μl of the enrichment broth from the growth mixture obtained from step 1 onto HBA medium containing 10 μg/ml of vancomycin; followed by incubation in similar growth conditions for 2–4 days [55].

The isolation of *C. concisus* from saliva samples can be achieved by streaking 6 μl of saliva on a HBA medium containing 10 μg/ml vancomycin and incubation under the above mentioned growth condition for 3 days. The mixed bacterial culture is then collected as a suspension in BHI broth and filtered using cellulose filter (pores size 0.65 μm) on a fresh HBA plate and incubated for 2 days [56]. However, this method might not reflect the original load of *C. concisus* in saliva due to potential further growth during both incubation periods. Furthermore, commensals that are resistant to vancomycin could also compete and reduce the growth of *C. concisus*.

5. Detection and confirmation by molecular methods

Historically, *C. concisus* has been identified based on conventional methods such as culturing. This technique poses many challenges and can provide false negative results due to several external factors. Molecular biology allows more reliability as well as a higher sensitivity when detecting the presence or absence of the pathogens. Therefore, a variety of molecular methods have been developed to detect *C. concisus*. For a bacterium of a fastidious nature, like *C. concisus*, molecular techniques can improve detection and identification in clinical samples. However, genetic variations should be considered in these detection methods otherwise some strains might be missed.

5.1 Direct detection of DNA in clinical samples

The presence of *C. concisus* DNA was investigated directly in faeces, intestinal biopsy, and saliva samples [11, 35, 56]. Initially a primer set (C412F and C1288R) designed by Linton et al. [22] to amplify the 16S rDNA gene (816 bp), was used as one step PCR to detect *C. concisus* from colonic biopsies [11]. Then, to identify *C. concisus*, the PCR product was sequenced and aligned to published sequences [11]. Soon after, a specific nested PCR was developed by Man et al. [40] to detect *C. concisus* in faecal specimens targeting the 16S rDNA gene. In the first PCR step, the primer set (C412F and C1288R) designed by Linton et al. [22], was used, while in the second step a new primer set (ConcisusF and ConcisusR) was developed to amplify a specific 560 bp region from the first PCR product of the 16S rDNA gene. Man et al. [40] applied this PCR to detect *C. concisus* from children's stool samples with CD, non-IBD patients and healthy controls. Later, this nested PCR method was also applied to detect *C. concisus* from saliva samples collected from IBD patients and healthy controls [56]. This nested PCR has been used for *C. concisus* DNA detection in other human clinical specimens including intestinal biopsies and saliva [40–42, 56].

Later on, Huq et al. [35] developed a multiplex PCR (m-PCR) to detect *C. concisus* and other *campylobacter* spp.. in faecal samples, based on the size of PCR product. When this m-PCR was applied on spiked faecal samples, *C. concisus, C. jejuni*, and *C. coli* were specifically identified at 10^5 cells/gm of faeces. However, as *C. concisus* is present in very low numbers in intestinal samples, using the nested PCR could be more sensitive than the m-PCR method.

5.2 Molecular confirmation and typing

The first specific PCR used for *C. concisus* identification was developed in 1995 by Bastyns et al. [57] using the forward primer MUC1 and a mix of two reverse primers CON1/CON2 to amplify the 23S rDNA region of *C. concisus* isolates. In 2004, we modified the this PCR method used by Bastyns et al. [57] to identify and group 19 clinical isolates from children with diarrhoea into two genomospecies using primers MUC1/CON1 (genomospecies A) and MUC1/CON2 (genomospecies B) [28]. However, there were some reports on that the primers designed for this PCR constantly cross reacted with *C. showae* and *Wolinella succinogenes* and produced a similar size PCR product [31].

Other techniques which successfully identified *C. concisus* were later developed including a two-step identification scheme for *Campylobacter, Arcobacter* and *Helicobacter* based on analysis of the 16S rRNA gene by PCR-RFLP (PCR-restriction fragment length polymorphism) by Marshall et al. [58]. Another PCR assay was developed, in 2001, from a 1.6 kb DNA fragment isolated from *C. concisus* genomic library for molecular identification, where a single PCR product was obtained without any cross reaction from other *Campylobacter* spp. [51]. Another primer set (Pcisus5-F and Pcisus6-R) developed by Matsheka et al. [51] was initially used to amplify DNA fragments (344 bp) obtained from a *C. concisus* genomic library and later showed to specifically amplify *C. concisus* [28, 35, 51]. This primer set amplifies *gyrB* [35].

6. Detection by MALDI-TOF

Matrix-assisted laser desorption/ionisation (MALDI) with time-of-flight mass spectrometry (TOFMS) is a technique developed more than three decades ago, which can be used to detect and characterise pathogens on the basis of larger biomolecules. Few studies proved the feasibility to identify *C. concisus* using MALDI-TOF-MS analysis of protein biomarkers from protein extracts of cell lysates or from whole cells [59–61]. The first attempt was in 2005 [62] using MALDI-TOF-MS to identify a number of *Campylobacter* species from their protein biomarkers, where they have identified a 10.5-kDa protein as the DNA-binding protein HU, and the potential species-identifying biomarker ions (SIBI) for *C. concisus* strains. Later this DNA-binding protein HU (10.5-kDa) was suggested to be used as a strain-specific biomarker for analysis by 'top-down' proteomics techniques [59]. However, a confirmed *C. concisus* isolate, by sequencing part of the 16S rRNA gene, could not be identified by MALDI-TOF by using database 3995 main spectra (June 2011) [63]. While a score ≥ 2.0 is considered reliable species identification, and between 1.7 and 2.0 represent reliable identification at the genus level, the isolate had only a score of 1.62, as *C. concisus* was not included in the database. The first successful identification and characterisation of *C. concisus* by MALDI-TOF and ClinProTools 2.2 software was in 2016 [60]. The study correctly identified all 14 *C. concisus* strains, despite evident differences between the isolates, with a scores ≥ 2.0 for secure species identification. There was a clear separation between other *Campylobacter* species and *C. concisus* by grouping of MSP dendrogram, with sufficient conserved peaks found for species identification. However, no distinguished biomarker has been identified to differentiate between the two genomospecies which can be easily differentiated by the 23S rDNA PCR [28]. Recently the lipo-oligosaccharide (LOS) structure (an important virulence factor which activates TLR4) of *C. concisus* clinical isolates correlated the inflammatory potential of each isolate with bacterial virulence by MALDI-TOF MS [61]. The presence of multiple bands in the SDS-PAGE profiles of *C. concisus* and *C. jejuni* LOS indicated their heterogeneity. The mass spectrometric analyses of lipid A indicated a novel hexa-acylated diglucosamine moiety, which cloud be an indicator of a potential virulence property.

7. Diversity of strains

As *C. concisus* is a genetically diverse organism, there is no standard molecular method yet to fully address this diversity. The standard typing of *C. concisus* could determine whether isolates obtained from diarrhoeic or IBD patients differ from those colonising healthy individuals [7, 64–66]. Applying such standard typing methods would help researchers to have a better understanding of *C. concisus* transmission, natural habitat, virulence and the host's immunological responses [67].

7.1 Typing by protein profiling techniques

It has been suggested that sodium dodecyl sulfate polyacrylamide gel electrophoresis (SDS-PAGE) and immunotyping were excellent tools for *C. concisus* identification. SDS-PAGE has been applied successfully to distinguish *C. concisus* from other small bacteria with very similar characteristics, such as *C. mucosalis*, other species of Gram-negative rods, as well as non-pigmenting and asaccharolytic bacteria [68, 69]. Other studies have used SDS-PAGE to identify clusters within *C. concisus;* the number of clusters identified varied from two to five [28, 65]. Therefore, the protein profiling technique could be discriminative for *C. concisus* isolates, but the discriminative power might be improved if combined with another typing technique such as genotyping. It should also be noted that protein profiling failed to separate *C. concisus* isolates from IBD patients and healthy controls [43].

7.2 Genomic typing by PCR and other techniques

There is no documented standard technique for genotyping *C. concisus* isolates. The first genetic method to type *C. concisus* was DNA–DNA hybridization [21]. The authors used electrophoretic protein profiles, immunotyping and DNA: DNA hybridization to identify 22 strains named as EF (E Falsen) group 22 which were identified as *C. concisus*. The 22 strains showed a considerable heterogeneity (42%) with the *C. concisus* type strain. The genetic diversity was later confirmed by analysis of 100 *C. concisus* isolates using randomly amplified polymorphic DNA (RAPD) [29, 70]. Another approach, which is a modification of the 23S rDNA PCR amplification method of Bastyns et al. [57], was used successfully used to type *C. concisus* by Istivan et al. using either the specific primer set (MUC1 and CON1) or (MUC1 and CON2) [28, 46]. In this system, isolates amplified by MUC1 and CON1 were assigned to genomospecies A while those amplified by MUC1 and CON2 were designated as genomospecies B [28, 66, 71].

The multi-locus sequence typing (MLST) was another technique applied to type 70 oral and intestinal *C. concisus* isolates from 8 patients with IBD and 6 healthy controls [43]. Subsequently, the neighbour-joining tree divided these isolates into 26 types and two major groups. Most isolates (87.5%) in cluster 1 were from IBD patients compared with only 28.6% in cluster 2 ($P < 0.05$). It was also reported that all of the invasive *C. concisus* isolates were localised in cluster 1 [43]. Two major groups were also demonstrated by MLST using a different set of housekeeping genes (*aspA*, *atpA*, *glnA*, *gltA*, *glyA*, *ilvD*, and *pgm*), applied to 60 *C. concisus* faecal isolates [72]. However, in both of these studies, it was not clear whether the two major groups were correlated with genomospecies A and B. A more recent study in 2016, Nielsen et al. [60] determined the genetic diversity of 67 *C. concisus* isolates from Danish diarrheic patients using MLST and specific differences in the 23S rRNA, and reported the high diversity of *C. concisus* with 53 sequence types (STs). However, dendrogram profiles of each allele showed a division into two groups, which was more or less correlated with genomospecies A and genomospecies B but had no association to the clinical severity of disease.

In addition to the previous techniques, the pulsed field gel electrophoresis (PFGE) indicated the diversity of *C. concisus* isolates and assigned them into two groups according to the source of the isolates (faeces and oral cavity) and into several subgroups [73]. Furthermore, *C. concisus* isolates have been allocated into four groups using the amplified fragment length polymorphism analysis technique (AFLP) [66, 71]. AFLP groups 1 and 2 aligned with genomospecies A and B (based on 23S rDNA PCR); while groups 3 and 4 could not be amplified by 23S rDNA PCR [66].

7.3 Typing by denaturing gradient gel electrophoresis (DGGE)

The PCR-DGGE technique was initially used to evaluate the microbial diversity in complex environments [5]. In environmental microbiology applications, universal primers are designed to target the 16S rDNA gene for the detection of mixed bacterial communities and differentiation of *Campylobacter*, *Helicobacter* and *Arcobacter* from clinical samples and *C. concisus* [5]. The PCR product is separated by polyacrylamide gel electrophoresis based on the use of different melting temperatures and its mobility in gradient denaturation of formamide and urea [74]. Previously, the DGGE technique was applied on 21 *C. jejuni* and one *C. coli* isolates using primer sets targeting the flagellin gene (fla-DGGE) [75]. A study conducted on DNA extracted from human saliva using PCR-DGGE to detect *Epsilobacteria*

(Campylobacters, Helicobacters and Arcobacters and related bacteria) reported that three reference *C. concisus* strains fell into two different DGGE profile groups [5]. However, Cornelius et al. [76] assigned *C. concisus* isolates from healthy volunteers and diarrhoea specimens into only one DGGE profile group by a semi-nested PCR-DGGE from 16S rRNA gene. The authors suggested that PCR-DGGE can be a useful tool for a direct detection of Epsiloproteobacteria.

Elshagmani [77] used Muyzer primer sets (518R: R-ATTACCGCGGCTGCTGG; 341F-GC: F-CCTACGGGAGGCAGCAG and 907R: R-CCGTCAATTCMT TTGAGTTT) [78, 79] to amplify the 16S DNA of *C. concisus*. Those primers were originally designed to detect and analyse the genetic diversity of mixed bacterial populations in environmental samples. As the DNA of all *C. concisus* tested isolates could be amplified using both Muyzer primer sets (1 and 2) to amplify 16S rDNA, it was suggested these sets could be used in clinical samples to detect *C. concisus* in mixed extracted DNA samples. Moreover, the analysis showed that all *Campylobacter* spp. isolates can be divided into four distinct groups that were defined as group I, II, III and IV. All *C. concisus* genomospecies B isolates fall into group II consistently, however, most genomospecies A isolates were allocated to group I but some were allocated in group II [77].

7.4 Typing by *rrn* analysis

The diversity of the ribosomal RNA (*rrn*) operon (5S rRNA, 16S rRNA, 23S rRNA genes, and the ITS regions) is considered a useful tool for differentiation of the heterogeneous *C. concisus* species [80]. The sequences of *C. concisus rrn* operons were used in a recent study for the purpose of strain typing and delineation of phylogenetic relationships within these operons. A total of 38 indels were identified in the *rrn* operon within *C. concisus* genome. Five indels found in the 23S rRNA gene were significantly associated with either genomospecies A or B ($p \leq 0.05$). The phylogenetic trees generated from 15 *rrn* operons and 23S rRNA genes also demonstrated sequence differences between strains within the *rrn*. Hence, the study confirmed that *C. concisus* can be classified into two genomospecies (A& B) based on the presence of the indels in the *rrn* operon and the 23S rRNA gene is a more reliable target for *C. concisus* typing than the 16S rRNA gene [80].

7.5 Whole genome sequencing

Whole-genome sequencing (WGS) is becoming increasingly available and affordable technique. Until 2011, there was only one *C. concisus* fully sequenced genome available for a strain (id. 13826) isolated from faeces of acute gastroenteritis patient and sequenced in 2007. The second *C. concisus* strain (UNSWCD) isolated from an intestinal biopsy of a patient with Crohn's disease was sequenced in 2011 [81]. Only 76% of genes were homologues between *C. concisus* 13826 and UNSWCD [82]. More *C. concisus* strains isolated from various clinical sources were sequenced and their genomes showed evidence of gene shuffling in *C. concisus* [83]. Few years later, another study defined the *C. concisus* core-genome and identified genomospecies-specific genes [84]. It concluded that the *C. concisus* core-genome, housekeeping genes, and the 23S rRNA gene consistently divided the 36 strains used in the study into two genomospecies. The study also reported novel genomic islands that contain type IV secretion system and putative effector proteins, in addition to other new genomic features. A study by our team investigated the *rrn* operon (5S rRNA, 16S rRNA, 23S rRNA genes, and the ITS regions) for four newly sequenced whole genomes extracted from intestinal and oral *C. concisus* strains, along with eight available WGSs online and established a clear correlation between the *rrn* operons and genomospecies [80].

More recently, the complete genome sequence of the *C. concisus* type strain ATCC 33237 and the draft genome sequences of eight additional well-characterised *C. concisus* strains were added to the database [85]. This was followed by a study in 2018 which analysed the genomes of 63 oral *C. concisus* strains isolated from patients with IBD and healthy controls, of which 38 genomes were newly sequenced. The genomes were examined to identify pathogenic molecular markers and the researchers reported a *C. concisus* molecular marker, which is a novel secreted enterotoxin B homologue (csep1-6bpi) potentially associated with active CD [86]. Moreover, in 2018 [87], a study to identify *C. concisus* virulence properties and adaptations capability to reside in the GI tract, produced robust genome sequencing data and comprehensive pangenome assessment from 53 new *C. concisus* strains. The researchers identified few genetic differences between oral and gut isolates from the same patient and suggested that the variability in bacterial secretion system content may play an important role in their virulence potential [87].

8. Conclusion

This chapter discussed the various approaches used to identify and differentiate *C. concisus*, since it was identified and named almost 40 years ago. This bacterium has been associated with periodontal diseases, acute enteritis, and IBDs, with the strongest evidence relating to acute and chronic intestinal diseases. However, its identification has always been challenging due to its inert biochemical characteristics and to the extremely high degree of genetic heterogeneity. The studies presented and explored in this chapter show that *C. concisus* is a genetically diverse species, but the extent of the difference between strains remains largely unknown. However, with the limitation of biochemical tests to identify *C. concisus*, molecular detection approaches including the PCR, of 23S rDNA, DGGE, m-PCR, MALDI-TOF and whole genome sequencing, have all made the identification and differentiation of this bacterium much easier than before.

Conflict of interest

There is no conflict of interest between the authors.

Author details

Mohsina Huq and Taghrid Istivan*
Biosciences and Food Technology, School of Science, RMIT University, Melbourne, Australia

*Address all correspondence to: Taghrid.istivan@rmit.edu.au

References

[1] Paster BJ, Gibbons RJ. Chemotactic response to formate by *Campylobacter concisus* and its potential role in gingival colonization. Infect Immun. 1986;52(2):378-83.

[2] Garrity GM, Brenner DJ, Krieg NR, Staley JT. Genus I. *Campylobacter*. Bergey's Manual of Syst Bacteriol. 2nd ed. New York: Springer; 2005.

[3] Tanner A, Badger S, Lai CH, Listgarten MA, Visconti RA, Socransky SS. *Wolinella* gen. nov., *Wolinella succinogenes* (*Vibrio succinogenes Wolin* et al.) comb. nov., and description of *Bateroides gracilis* sp.nov., *Wolinella recta* sp. nov., *Camylobacter concisus* sp. nov., and *Eikenella corrodens* from humans with periodontal disease. Int J Syst Bacteriol. 1981(31):432-45.

[4] Scanlon KA, Cagney C, Walsh D, McNulty D, Carroll A, McNamara EB, et al. Occurrence and characteristics of fastidious *Campylobacteraceae* species in porcine samples. Int J Food Microbiol. 2013;163(1):6-13. doi: 10.1016/j.ijfoodmicro.2013.02.004.

[5] Petersen RF, Harrington CS, Kortegaard HE, On SL. A PCR-DGGE method for detection and identification of *Campylobacter, Helicobacter, Arcobacter* and related Epsilobacteria and its application to saliva samples from humans and domestic pets. J Appl Microbiol. 2007;103(6):2601-15. doi: 10.1111/j.1365-2672.2007.03515.x.

[6] Chaban B, Ngeleka M, Hill JE. Detection and quantification of 14 *Campylobacter* species in pet dogs reveals an increase in species richness in feces of diarrheic animals. BMC Microbiol. 2010;10:73. doi: 1471-2180-10-73 [pii] 10.1186/1471-2180-10-73.

[7] Macuch PJ, Tanner AC. *Campylobacter* species in health, gingivitis, and periodontitis. J Dent Res. 2000;79(2):785-92.

[8] Kamma JJ, Diamanti-Kipioti A, Nakou M, Mitsis FJ. Profile of subgingival microbiota in children with primary dentition. J Periodontal Res. 2000;35(1):33-41.

[9] Kamma JJ, Diamanti-Kipioti A, Nakou M, Mitsis FJ. Profile of subgingival microbiota in children with mixed dentition. Oral Microbiol Immunol. 2000;15(2):103-11.

[10] Dewhirst FE, Chen T, Izard J, Paster BJ, Tanner AC, Yu WH, et al. The human oral microbiome. J Bacteriol. 2010;192(19):5002-17. doi: 10.1128/JB.00542-10.

[11] Zhang L, Man SM, Day AS, Leach ST, Lemberg DA, Dutt S, et al. Detection and isolation of *Campylobacter* species other than *C. jejuni* from children with Crohn's disease. J Clin Microbiol. 2009;47(2):453-5. doi: JCM.01949-08 [pii] 10.1128/JCM.01949-08.

[12] Tanner AC, Dzink JL, Ebersole JL, Socransky SS. *Wolinella recta, campylobacter concisus, Bacteroides gracilis*, and *Eikenella corrodens* from periodontal lesions. J Periodont Res. 1987;22(4):327-30.

[13] HaffajeeAD,SocranskySS,EbersoleJL, Smith DJ. Clinical, microbiological and immunological features associated with the treatment of active periodontosis lesions. J Clin Periodontol. 1984;11(9):600-18.

[14] Taubman MA, Haffajee AD, Socransky SS, Smith DJ, Ebersole JL. Longitudinal monitoring of humoral antibody in subjects with destructive periodontal diseases. J Periodontal Res. 1992;27(5):511-21.

[15] Tanner A, Maiden MF, Macuch PJ, Murray LL, Kent RL, Jr. Microbiota of health, gingivitis, and initial

periodontitis. J Clin Periodontol. 1998;25(2):85-98.

[16] Kamma JJ, Nakou M, Manti FA. Microbiota of rapidly progressive periodontitis lesions in association with clinical parameters. J Periodontol. 1994;65(11):1073-8.

[17] Kamma JJ, Lygidakis NA, Nakou M. Subgingival microflora and treatment in prepubertal periodontitis associated with chronic idiopathic neutropenia. J Clin Periodontol. 1998;25(9):759-65.

[18] Kamma JJ, Nakou M, Baehni PC. Clinical and microbiological characteristics of smokers with early onset periodontitis. J Periodontal Res. 1999;34(1):25-33.

[19] Kamma JJ, Nakou M, Persson RG. Association of early onset periodontitis microbiota with aspartate aminotransferase activity in gingival crevicular fluid. J Clin Periodontol. 2001;28(12):1096-105.

[20] Socransky SS, Haffajee AD. Periodontal microbial ecology. Periodontol 2000. 2005;38:135-87. doi: 10.1111/j.1600-0757.2005.00107.x.

[21] Vandamme P, Falsen E, Pot B, Hoste B, Kersters K, De Ley J. Identification of EF group 22 campylobacters from gastroenteritis cases as *Campylobacter concisus*. J Clin Microbiol. 1989;27(8):1775-81.

[22] Linton D, Owen RJ, Stanley J. Rapid identification by PCR of the genus *Campylobacter* and of five *Campylobacter* species enteropathogenic for man and animals. Res Microbiol. 1996;147(9):707-18. doi: S0923-2508(97)85118-2 [pii].

[23] Kulkarni SP, Lever S, Logan JM, Lawson AJ, Stanley J, Shafi MS. Detection of *campylobacter* species: a comparison of culture and polymerase chain reaction based methods. J Clin Pathol. 2002;55(10):749-53.

[24] Samie A, Obi CL, Barrett LJ, Powell SM, Guerrant RL. Prevalence of *Campylobacter* species, *Helicobacter pylori* and *Arcobacter* species in stool samples from the Venda region, Limpopo, South Africa: studies using molecular diagnostic methods. J Infect. 2007;54(6):558-66. doi: S0163-4453(06)00377-X [pii] 10.1016/j.jinf.2006.10.047.

[25] Lauwers s, Devreker T, Van Etterijck R. Isolation of *Campylobacter Concisus* from human faeces. *Microb Ecol Health Dis* 4 (suppl). 1991:991.

[26] Lindblom GB, Sjogren E, Hansson-Westerberg J, Kaijser B. *Campylobacter upsaliensis, C. sputorum sputorum* and *C. concisus* as common causes of diarrhoea in Swedish children. Scand J Infect Dis. 1995;27(2):187-8.

[27] Russell J. *Campylobacter* like organisms: Investigation of clinical and phenotypical aspects. Applied Biology and Biotechnology. Melbourne: RMIT University; 1995. p. 15-41.

[28] Istivan TS, Coloe PJ, Fry BN, Ward P, Smith SC. Characterization of a haemolytic phospholipase A(2) activity in clinical isolates of *Campylobacter concisus*. J Med Microbiol. 2004;53(Pt 6):483-93.

[29] Van Etterijck R, Breynaert J, Revets H, Devreker T, Vandenplas Y, Vandamme P, et al. Isolation of *Campylobacter concisus* from feces of children with and without diarrhea. J Clin Microbiol. 1996;34(9):2304-6.

[30] Musmanno RA, Russi M, Figura N, Guglielmetti P, Zanchi A, Signori R, et al. Unusual species of campylobacters isolated in the Siena Tuscany area, Italy. New Microbiol. 1998;21(1):15-22.

[31] Engberg J, On SL, Harrington CS, Gerner-Smidt P. Prevalence of *Campylobacter, Arcobacter, Helicobacter,* and *Sutterella* spp. in human fecal samples as estimated by a reevaluation of isolation methods for Campylobacters. J Clin Microbiol. 2000;38(1):286-91.

[32] Lastovica AJ, le Roux E. Efficient isolation of campylobacteria from stools. J Clin Microbiol. 2000;38(7):2798-9.

[33] Maher M, Finnegan C, Collins E, Ward B, Carroll C, Cormican M. Evaluation of culture methods and a DNA probe-based PCR assay for detection of *Campylobacter* species in clinical specimens of feces. J Clin Microbiol. 2003;41(7):2980-6.

[34] Vandenberg O, Cornelius AJ, Souayah H, Martiny D, Vlaes L, Brandt SM, et al. The role of Epsilonproteobacteria in children with gastroenteritis. Pediatr Infect Dis J. 2013;32(10):1140-2. doi: 10.1097/INF.0b013e3182975047.

[35] Huq M, Gonis G, Istivan T. Development and Evaluation of a Multiplex PCR for the Detection of *Campylobacter concisus* and Other *Campylobacter* spp. from Gastroenteritis Cases. Open J Med Microbiol. 2014;4(1). doi: 10.4236/ojmm.2014.41005

[36] Underwood AP, Kaakoush NO, Sodhi N, Merif J, Lee WS, Riordan SM, et al. *Campylobacter concisus* pathotypes are present at significant levels in patients with gastroenteritis. J Med Microbiol. 2015;65(3):219-26. doi: 10.1099/jmm.0.000216.

[37] Nielsen HL, Ejlertsen T, Engberg J, Nielsen H. High incidence of *Campylobacter concisus* in gastroenteritis in North Jutland, Denmark: a population-based study. Clin Microbiol Infect. 2013;19(5):445-50. doi: 10.1111/j.1469-0691.2012.03852.x.

[38] Serichantalergs O, Ruekit S, Pandey P, Anuras S, Mason C, Bodhidatta L, et al. Incidence of *Campylobacter concisus* and *C. ureolyticus* in traveler's diarrhea cases and asymptomatic controls in Nepal and Thailand. Gut pathogens. 2017;9:47-. doi: 10.1186/s13099-017-0197-6.

[39] Aabenhus R, Hynes SO, Permin H, Moran AP, Andersen LP. Lectin typing of *Campylobacter concisus*. J Clin Microbiol. 2002;40(2):715-7.

[40] Man SM, Zhang L, Day AS, Leach ST, Lemberg DA, Mitchell H. *Campylobacter concisus* and other *Campylobacter* species in children with newly diagnosed Crohn's disease. Inflamm Bowel Dis. 2010;16(6):1008-16. doi: 10.1002/ibd.21157.

[41] Mukhopadhya I, Thomson JM, Hansen R, Berry SH, El-Omar EM, Hold GL. Detection of *Campylobacter concisus* and other *Campylobacter* species in colonic biopsies from adults with ulcerative colitis. PLoS One. 2011;6(6):e21490. doi: 10.1371/journal.pone.0021490 PONE-D-11-06834 [pii].

[42] Mahendran V, Riordan SM, Grimm MC, Tran TA, Major J, Kaakoush NO, et al. Prevalence of *Campylobacter* species in adult Crohn's disease and the preferential colonization sites of *Campylobacter* species in the human intestine. PLoS One. 2011;6(9):e25417. doi: 10.1371/journal.pone.0025417 PONE-D-11-15438 [pii].

[43] Ismail Y, Mahendran V, Octavia S, Day AS, Riordan SM, Grimm MC, et al. Investigation of the enteric pathogenic potential of oral *Campylobacter concisus* strains isolated from patients with inflammatory bowel disease. PLoS One. 2012;7(5):e38217. doi: 10.1371/journal.pone.0038217.

[44] Kaakoush NO, Castano-Rodriguez N, Day AS,

Lemberg DA, Leach ST, Mitchell HM. *Campylobacter concisus* and exotoxin 9 levels in paediatric patients with Crohn's disease and their association with the intestinal microbiota. J Med Microbiol. 2014;63(Pt 1):99-105. doi: 10.1099/jmm.0.067231-0.

[45] Zhang L, Lee H, Grimm MC, Riordan SM, Day AS, Lemberg DA. *Campylobacter concisus* and inflammatory bowel disease. World J Gastroenterol. 2014;20(5):1259-67. doi: 10.3748/wjg.v20.i5.1259.

[46] Istivan T, Peter W, Peter C. *Campylobacter concisus*: An emerging pathogen of the gastrointestinal tract. Current Research, Technology and Education Topics in Applied Microbiology and Microbial Biotechnology. 2010:626-34.

[47] Lee H, Ma R, Grimm MC, Riordan SM, Lan R, Zhong L, et al. Examination of the anaerobic growth of *Campylobacter concisus* strains. Int J Microbiol. 2014;2014:476047. doi: 10.1155/2014/476047.

[48] On SL. Confirmation of human *Campylobacter concisus* isolates misidentified as *Campylobacter mucosalis* and suggestions for improved differentiation between the two species. J Clin Microbiol. 1994;32(9):2305-6.

[49] Lastovica AJ, Le Roux E, Warren R, Klump H. Additional data on clinical isolates of *Campylobacter mucosalis*. J Clin Microbiol. 1994;32(9):2338-9.

[50] Bolton FJ, Hutchinson DN, Parker G. Reassessment of selective agars and filtration techniques for isolation of *Campylobacter* species from faeces. Eur J Clin Microbiol Infect Dis. 1988;7(2):155-60.

[51] Matsheka MI, Lastovica AJ, Elisha BG. Molecular identification of *Campylobacter concisus*. J Clin Microbiol. 2001;39(10):3684-9. doi: 10.1128/JCM.39.10.3684-3689.2001.

[52] Nielsen HL, Engberg J, Ejlertsen T, Nielsen H. Comparison of polycarbonate and cellulose acetate membrane filters for isolation of *Campylobacter concisus* from stool samples. Diagn Microbiol Infect Dis. 2013;76(4):549-50. doi: 10.1016/j.diagmicrobio.2013.05.002.

[53] Blaser MJ, NachamkinI,SzymanskiCM,NachamkinI, SzymanskiCM, BlaserMJ. *Campylobacter*. 3rd ed. Washington, DC: ASM Press; 2008.

[54] Lastovica AJ, On SLW, Zhang L. The Family *Campylobacteraceae*. In: Rosenberg E, DeLong EF, Lory S, Stackebrandt E, Thompson F, editors. The Prokaryotes: Deltaproteobacteria and Epsilonproteobacteria. Berlin, Heidelberg: Springer Berlin Heidelberg; 2014. p. 307-35.

[55] Kaakoush NO, Deshpande NP, Wilkins MRT, C. G. , Burgos-Portugal JA, Raftery MJ, Day AS, et al. The pathogenic potential of *Campylobacter concisus* strains associated with chronic intestinal diseases. PLoS One. 2011;6(12):e29045. doi: 10.1371/journal.pone.0029045 PONE-D-11-18302 [pii].

[56] Zhang L, Budiman V, Day AS, Mitchell H, Lemberg DA, Riordan SM, et al. Isolation and detection of *Campylobacter concisus* from saliva of healthy individuals and patients with inflammatory bowel disease. J Clin Microbiol. 2010;48(8):2965-7. doi: JCM.02391-09 [pii] 10.1128/JCM.02391-09.

[57] Bastyns K, Chapelle S, Vandamme P, Goossens H, De Wachter R. Specific detection of *Campylobacter concisus* by PCR amplification of 23S rDNA areas. Mol Cell Probes. 1995;9(4):247-50. doi: S0890-8508(95)90114-0 [pii].

[58] Marshall SM, Melito PL, Woodward DL, Johnson WM, Rodgers FG, Mulvey MR. Rapid identification of *Campylobacter, Arcobacter,* and *Helicobacter* isolates by PCR-restriction fragment length polymorphism analysis of the 16S rRNA gene. J Clin Microbiol. 1999;37(12):4158-60.

[59] Fagerquist CK, Yee E, Miller WG. Composite sequence proteomic analysis of protein biomarkers of *Campylobacter coli, C. lari* and *C. concisus* for bacterial identification. Analyst. 2007;132(10):1010-23. doi: 10.1039/b702859h.

[60] Nielsen HL, Mølvadgaard M, Nielsen H, M K. Identification and Differentiation of Highly Diverse *Campylobacter concisus* Strains using the MALDI Biotyper. Clin Microbial 2016;5(1):230. doi: 10.4172/2327-5073.1000230.

[61] Brunner K, John CM, Phillips NJ, Alber DG, Gemmell MR, Hansen R, et al. Novel *Campylobacter concisus* lipooligosaccharide is a determinant of inflammatory potential and virulence. J Lipid Res. 2018;59(10):1893-905. doi: 10.1194/jlr.M085860.

[62] Mandrell RE, Harden LA, Bates A, Miller WG, Haddon WF, Fagerquist CK. Speciation of *Campylobacter coli, C. jejuni, C. helveticus, C. lari, C. sputorum,* and *C. upsaliensis* by Matrix-Assisted Laser Desorption Ionization–Time of Flight Mass Spectrometry. Applied and Environmental Microbiology. 2005;71(10):6292-307. doi: 10.1128/AEM.71.10.6292-6307.2005%J Applied and Environmental Microbiology.

[63] Hess DL, Pettersson AM, Rijnsburger MC, Herbrink P, van den Berg HP, Ang CW. Gastroenteritis caused by *Campylobacter concisus*. J Med Microbiol. 2012;61(Pt 5):746-9. doi: 10.1099/jmm.0.032466-0.

[64] Newell DG. *Campylobacter concisus*: an emerging pathogen? Eur J Gastroenterol Hepatol. 2005;17(10):1013-4. doi: 10.1097/00042737-200510000-00001.

[65] Engberg J, Bang DD, Aabenhus R, Aarestrup FM, Fussing V, Gerner-Smidt P. *Campylobacter concisus*: an evaluation of certain phenotypic and genotypic characteristics. Clin Microbiol Infect. 2005;11(4):288-95. doi: CLM1111 [pii] 10.1111/j.1469-0691.2005.01111.x.

[66] Aabenhus R, On SL, Siemer BL, Permin H, Andersen LP. Delineation of *Campylobacter concisus* genomospecies by amplified fragment length polymorphism analysis and correlation of results with clinical data. J Clin Microbiol. 2005;43(10):5091-6. doi: 43/10/5091 [pii] 10.1128/JCM.43.10.5091-5096.2005.

[67] Lastovica A. Emerging *Campylobacter* spp.: the Tip of the Iceberg. Clin Microbiol Newsl. 2006;28(7):49-55.

[68] Tanner AC. Characterization of *Wolinella* spp., *Campylobacter concisus, Bacteroides gracilis,* and *Eikenella corrodens* by polyacrylamide gel electrophoresis. J Clin Microbiol. 1986;24(4):562-5.

[69] Penner JL. The genus *Campylobacter*: a decade of progress. Clin Microbiol Rev. 1988;1(2):157-72. doi: 10.1128/cmr.1.2.157.

[70] Matsheka MI, Lastovica AJ, Zappe H, Elisha BG. The use of (GTG)5 oligonucleotide as an RAPD primer to type *Campylobacter concisus*. Lett Appl Microbiol. 2006;42(6):600-5. doi: LAM1900 [pii] 10.1111/j.1472-765X.2006.01900.x.

[71] Kalischuk LD, Inglis GD. Comparative genotypic and pathogenic examination of *Campylobacter concisus* isolates from diarrheic and

non-diarrheic humans. BMC Microbiol. 2011;11:53. doi: 1471-2180-11-53 [pii] 10.1186/1471-2180-11-53.

[72] Miller WG, Chapman MH, Yee E, On SL, McNulty DK, Lastovica AJ, et al. Multilocus sequence typing methods for the emerging *Campylobacter* Species *C. hyointestinalis, C. lanienae, C. sputorum, C. concisus,* and *C. curvus*. Front Cell Infect Microbiol. 2012;2:45. doi: 10.3389/fcimb.2012.00045.

[73] Matsheka MI, Elisha BG, Lastovica AL, On SL. Genetic heterogeneity of *Campylobacter concisus* determined by pulsed field gel electrophoresis-based macrorestriction profiling. FEMS Microbiol Lett. 2002;211(1):17-22.

[74] Fischer SG, Lerman LS. DNA fragments differing by single base-pair substitutions are separated in denaturing gradient gels: correspondence with melting theory. Proc Natl Acad Sci U S A. 1983;80(6):1579-83. doi: 10.1073/pnas.80.6.1579.

[75] Najdenski H, Heyndrickx M, Herman L, Messens W. Fla-DGGE analysis of *Campylobacter jejuni* and *Campylobacter coli* in cecal samples of broilers without cultivation. Vet Microbiol. 2008;127(1-2):196-202. doi: 10.1016/j.vetmic.2007.08.002.

[76] Cornelius AJ, Chambers S, Aitken J, Brandt SM, Horn B, On SL. Epsilonproteobacteria in humans, New Zealand. Emerg Infect Dis. 2012;18(3):510-2. doi: 10.3201/eid1803.110875.

[77] Elshagmani E. The role of *Campylobacter concisus* in enteric infections. School of Applied Sciences,. Melbourne, Australia: RMIT University, Melbourne, Australia; 2015.

[78] Muyzer G, de Waal EC, Uitterlinden AG. Profiling of complex microbial populations by denaturing gradient gel electrophoresis analysis of polymerase chain reaction-amplified genes coding for 16S rRNA. Appl Environ Microbiol. 1993;59(3):695-700.

[79] Muyzer G, Smalla K. Application of denaturing gradient gel electrophoresis (DGGE) and temperature gradient gel electrophoresis (TGGE) in microbial ecology. Antonie Van Leeuwenhoek. 1998;73(1):127-41. doi: 10.1023/a:1000669317571.

[80] Huq M, Van TTH, Gurtler V, Elshagmani E, Allemailem KS, Smooker PM, et al. The ribosomal RNA operon (*rrn*) of *Campylobacter concisus* supports molecular typing to genomospecies level. Gene Reports. 2017;6:8-14. doi: http://dx.doi.org/10.1016/j.genrep.2016.10.008.

[81] Deshpande NP, Kaakoush NO, Mitchell H, Janitz K, Raftery MJ, Li SS, et al. Sequencing and validation of the genome of a *Campylobacter concisus* reveals intra-species diversity. PLoS One. 2011;6(7):e22170. doi: 10.1371/journal.pone.0022170 PONE-D-11-06842 [pii].

[82] Kaakoush NO, Deshpande NP, Wilkins MR, Raftery MJ, Janitz K, Mitchell H. Comparative analyses of *Campylobacter concisus* strains reveal the genome of the reference strain BAA-1457 is not representative of the species. Gut Pathog. 2011;3:15. doi: 1757-4749-3-15 [pii] 10.1186/1757-4749-3-15.

[83] Deshpande NP, Kaakoush NO, Wilkins MR, Mitchell HM. Comparative genomics of *Campylobacter concisus* isolates reveals genetic diversity and provides insights into disease association. BMC Genomics. 2013;14:585. doi: 10.1186/1471-2164-14-585.

[84] Chung HK, Tay A, Octavia S, Chen J, Liu F, Ma R, et al. Genome

analysis of *Campylobacter concisus* strains from patients with inflammatory bowel disease and gastroenteritis provides new insights into pathogenicity. Sci Rep. 2016;6:38442. doi: 10.1038/srep38442.

[85] Cornelius AJ, Miller WG, Lastovica AJ, On SLW, French NP, Vandenberg O, et al. Complete Genome Sequence of *Campylobacter concisus* ATCC 33237(T) and Draft Genome Sequences for an Additional Eight Well-Characterized *C. concisus* Strains. Genome announcements. 2017;5(29):e00711-17. doi: 10.1128/genomeA.00711-17.

[86] Liu F, Ma R, Tay CYA, Octavia S, Lan R, Chung HKL, et al. Genomic analysis of oral *Campylobacter concisus* strains identified a potential bacterial molecular marker associated with active Crohn's disease. Emerging microbes & infections. 2018;7(1):64-. doi: 10.1038/s41426-018-0065-6.

[87] Gemmell MR, Berry S, Mukhopadhya I, Hansen R, Nielsen HL, Bajaj-Elliott M, et al. Comparative genomics of *Campylobacter concisus*: Analysis of clinical strains reveals genome diversity and pathogenic potential. Emerging Microbes & Infections. 2018;7(1):1-17. doi: 10.1038/s41426-018-0118-x.

Chapter 6

Health Care Associated Infections (HCAIs) a New Threat for World; U-Turn from Recovery to Death

Ayesha Noor, Ali Raza Ishaq, Laila Jafri, Faiza Jabeen, Rehana Rani, Bushra Hafeez Kiani, Nosheen Akhtar, Zeeshan Javed, Tahira Younis and Fatima Jalal

Abstract

Health care associated infections also termed as nosocomial infections are notable cause of morbidity and mortality especially in resource limited countries like Pakistan. Newborns and aged people have more probability of being infected by Health care associated infections because of immunosuppressant. Central line associated blood stream infections (CLABSI) are considered as one of the promising negotiator associated with Health Care associated infections. Improper health care setting and unaware medical staff play a championship protagonist in prevalence of health care associated infections. Standard hygienic measures should be adopted to reduce risk of Health care associated infections. So, there is a pressing need to take on control policies by Government to handle this dilemma. This chapter gives new intuition to healthcare associated microbes, infections and provides comprehensive detailed on ironic precaution to scientific community.

Keywords: Palindromic rheumatism, Rheumatoid arthritis, Environmental risk factors, Genetic risk factors, Therapies

1. Introduction

In health care safety issues, health care associated infections (HCAIs) are a significant cause of morbidity and mortality in developing countries specially in Pakistan. Environment of hospital favors certain infections during the period of admission patients, these are termed as Health care associated infections. Contaminated equipment's, unaware medical staff, unhealthy hospital environment and not satisfactory standard measures promote Hospital acquire infections, nosocomial infections/Health care associated infections (HCAIs). Prevalence of health care associated infection is high in developing countries due to unhealthy health care settings, where it affects more than 25–30% patients. Unhealthy Standard hygienic measures and risk of HCAIs are directly related which clearly address a pressing need to follow standard hygienic guidelines [1–3]. Prevalence of HCAIs is roughly about 10–30% in developing countries and 5–10% in developed countries [4].

IntechOpen

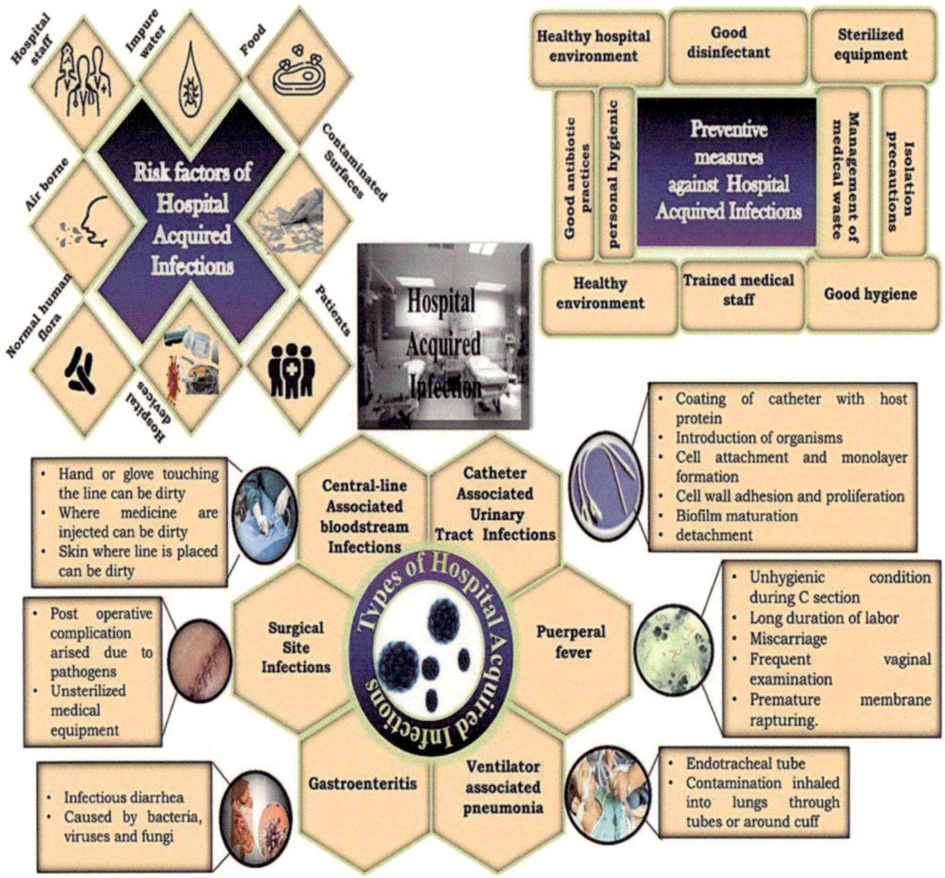

Figure 1.
Overview of hospital acquired infections.

Of all hospitalized patients about 15% are infected by HCAIs. In developing countries ten in every hundred acquire HCAIs. Neonates, patients of organ transplant, patients of burn surgery and patients at Intensive Care Unit (ICU) are more prone to HCAIs. High rate of infection is observed in ICU ward. HCAIs not only down health of patients who are already ill with other diseases but also impose socio economic burden for developing countries by increasing health care cost [5, 6]. This review article not only addresses endemic threat for patients but also covers counter measures to handle this problem as shown in **Figure 1**.

Role of medicines in treatment of diseases is understood by many of us but in recent years awareness about communication of diseases through health care is increased [7]. Ignaz Phillip Semmelweis was medical doctor who realized communication of puerperal sepsis through hospitals. He found increased rate of women death in clinic. To reduce rate of maternal deaths Semmelweis introduced chlorinated lime for hand washing. It is estimated that about 100,000 people are killing in world through HCAIs [8].

This review paper provides us advances knowledge about hospital infections and provide instruction to government for the improvement of medical conditions. In this paper, we have summarized the various health care associated infections by which patients are more vulnerable ultimately conditions will be severe.

2. Types of HCAIs

2.1 Central line-associated bloodstream infections (CLABSI)

Central line-association bloodstream infections are the infections of central venous catheter (CVC) by which catheter tube is only route for microorganisms to enter patient body and central line is required for infections progression within 48 hour [9]. There are two major routes which are adopted by microorganism, intraluminal and extraluminal but intraluminal route is important in the sense of causing severity through catheter hub [10]. In above luminal routes, biofilms are formed due catheter hub infection, composed of bacteria which is formed mostly extracellular matrix within 24 hour of catheter insertion [11]. CLABS-infections leads to cancers and other neurodegenerative diseases in those patients which are immunosuppressant [12], agonize chemotherapy and confess in ICU. Central-Line Insertion [13] is best method to control CLABSIs in ICU patients, but it is very cost effective.

CLABSIs effect the neonatal life, in the form of sepsis which cause the 20–36% [14] due to CVCs. In most cases, the babies which are premature, exposed to CLABSIs, have poor growth [15], high death rate and neurodegenerative diseases. Fever, hypothermia, apnea and bradycardia are most disastrous indicators which are appeared in the 1 year < age children [16]. CLABSIs also effect the adult life, the proportion of gram-negative bacteria including *Klebsiella pneumoniae, E. coli, and Enterobacter cloacae,* exceeded [17] due to translocation from gastrointestinal tract which enhances the bowel wall infections and mucosa infections. All these microorganisms enter the human to mutate the normal body functioning.

The potential pathway for source of microorganism are following catheter insertion site, hands of healthcare worker, contaminated disinfected, patient skin flora during catheter hub operation, contaminated drugs or fluids, catheter infections and hematogenous dissemination as a secondary infections [18]. Through these routes, microorganism enters in body and form biofilm at insertion site. Biofilms contains colony of bacteria which is formed firstly extracellular at catheter site but with a passage of time, move towards intracellular matrix [19]. The pathophysiological features are low metabolic rate, tiniest inhibitory concentration, less vulnerable to antibiotics and high penetrance rate to antibiotic, adapted by microorganism to spread the infections [20].

According recent researches there are 84,000-204,000 [21] people, infected by CLABSIs and 25,000 death. Death incidence rate of CLABSI is about 12–25%. Catheters are used for intravenous therapies, for delivery of specific medicine and specific treatment. Through contaminated infusion of catheter and unhealthy environment microbes gain access to bloodstream which cause CLABSI. Coagulase negative staphylococci for example *Staphylococcus epidermidis*is and *S. aureus* [22] which are most common cause of CLABSI in developing countries including south Korea [23]. According to a study of intensive care unit in Pakistan CLABSI has highest incidence rate in all HCAIs [24].

To overcome such rate, government and hospital admin should adopt following aliments, sterile barrier [25] are used during catheter insertion and use disinfectants in case of intravenous administration [26]. Government should also give priority the potent disinfectants including's, chlorhexidine, povidone iodine, iodophor and 70% alcohol but optimal timing is unclear yet [9]. Awareness in medical staff, PICC site assortment, CVC insertion and maintenance by intervention bundles, applications skin antiseptics, In-line filters, umbilical catheter, catheter dressing, prophylactic antimicrobial and antimicrobial locks are the methods of preventions for CLABSI.

Awareness in medical staff Basic principle to control any disturbance, educate the people to specific issue. CLABSI incidence rate would be higher day by day than the responsibility of government and hospital admin to educate their staff [27]. Government should also publish proper set of rules and guidelines for maintained and insertion of catheter, insist the staff to follow these rules and do more practice [28]. The potential way to prevent CLABSI, to enrolled only skilled nurse for insertion and upheld of catheter and nurse-to-patient ratio should kept normal specifically in ICU [9].

PICC site assortment Upper and lower limb [29] considered as PICC site but the exact location is unknown yet. Subclavian and femoral vein [30] are the most suitable site for insertion as compare to jugular vein because it is more susceptible for infections and biofilms formation [28].

CVC insertion and maintenance by intervention bundles Intervention bundles (IBs) are widely used to control bloodstream infections and maintenance of CVC but before introducing the IBs, medical staff must be monitored checklist to reassure compliance and the recommendations. Recent health care report revealed that 40% CLABSI patient abridged by using intervention bundles in USA [31].

Applications of skin antiseptics Skin antiseptics stunt the growth of microorganisms on living tissues, 70% alcohol, tincture of iodine or alcoholic chlorhexidine gluconate (CHG)solutions are widely used before inserting catheter [32]. Researchers proved that there is no recommendation for preference or valuable any antiseptics among others but precautions for usage of antiseptic should be followed such as antiseptic must be dried up before inserting catheter and changing the dressing [33]. Most important thing, application of antiseptic (CUG Solution) on infants could lead to skin cancer and neurological disorder. Overindulge of iodine tincture metamorphose the functional veracity of thyroid gland which need iodine to release thyroxin [34, 35].

In-line filter, a device which is used to pour the material into body but not to prevent the CLABSI. According to reported data, there are two inline filters used such 0.2 μ and 1.2 μ used for liquid and large molecule insertion respectively. In-line filter basically reduced the mortality and morbidity in neonates [36].

Umbilical Catheter It is used for monitoring the sickness in neonates through arterial and venous umbilical catheter. So, before inserting the catheter on umbilical site, antiseptic must be applied to prevent complications. Despite of these, antibiotics especially low dose of heparin is also used to control CLABSIs. Optimal time period for catheter is 5 to 14 days either for venous or arteriosus [9].

Catheter dressing including gauze and transparent clothing are mostly used in CLABSI-site until used when the bleeding or oozing is not stunt. But researcher endorsed that antibiotic is not a helpful font to stop these type of infection because antibiotics induced the fungal infections and resist the bacterial at catheter insertion site [9].

Prophylactic antimicrobial is most effective agent to stunt the mortality rate in newborn and resist the microorganisms but the exact mechanism for usage of agent on CLABSI is not clear [37].

Antimicrobial locks including fusidic acid, vancomycin and amikacin are most effective agents to prevent the CLABSI specifically in newborn babies [38].

2.2 Catheter associated urinary tract infections (CAUTI)

Catheterization is a process of introducing urinary catheter into urinary bladder which functions both as therapeutic tool and diagnostic tool. In health care facilities catheter associated urinary tract infections (UTIs) are most common infection. Infection can occur during insertion of catheter and cleaning of catheter if process is done inadvertently. For number of reasons about 25% of all hospitalized patients need catheter and risk of catheter associated urinary tract infection is much higher in Intensive Care Unit (ICU). Among all HCAIs catheter associated urinary tract infection account for about 40% [39, 40], catheter related UTIs 70% and 95% UTIs in intensive care units [41]. To reduce risk of catheter associated urinary tract infections there is a pressing need to follow standard measures during catheterization process and safe maintenance of catheterization. If Catheter associated urinary tract infections are ignored for long time serious kidney disorders may arise [42, 43].

Pandemic nature of CAUTIs, 150 million affect the people annually which show following symptoms such as somber sequelae, recurrences, pyelonephritis with sepsis, blood with urine, catheter obstruction and renal damage [44]. Accounted symptoms are the result of severe complex metabolic reactions due to overdose of antibiotic, frequently usage of antimicrobial drugs such as *Clostridium difficile* colitis.3 [45]. There are two major category of UTIs, complicated UTIs and Uncomplicated UTIs based on pathophysiological complication [46]. In case of uncomplicated UTIs are also known as community-onset-cystitis [47] in which patient remain healthy, not develop any neurological problems of urinary system [48]. These types of complications mostly recorded in female, but infant or older men could be exposed for UTIs [49, 50].

Recent discoveries proved that complicated UTIs totally dependent on physiological pools of patients [51]. If a person has weak immune defense system, renal failure, renal stones, urodynamics and indwelling catherization (IC), are major indicator for UTIS but the IC is most communal agents to progress the infections [52].

E. coli is most common cause of CAUTIs but other are listed in graph:1. These microorganisms progress the infections after 24–48 hours of catheter insertion. Entrance of microorganisms form the biofilms inside the catheter-site, which prevent the action of antibiotics but permit the microorganism to inside the patient body [52]. Microbes divide rapidly to develop infections, patient suffered 3–7% microbes daily after catheter-insertion [53]. These microbes mostly gram-negative bacteria which cause CAUTIs, enter the urinary system via crossing periurethral area [44]. The gram-negative bacteria are potential reservoir of infections, patients of CAUTIs are epidemic in nature so these gram-negative bacteria have efficient resistant again antimicrobial therapies [54]. If a patient remains untreated than this disease become acute [53].

This infection can be diagnosed by urinalysis test which address the presence of leucocytes and nitrites in urine but not detach these compound. The presence of leucocytes and nitrites signpost that a person is suffered from CAUTIs and progression of infections. Leucocytes in urine are the result of, activation of leucocytes esterase (LE), which is immune system product, triggering the malformed and break down of normal WBCs through the action of microorganisms. But the presence of nitrites, developed curiosity in nitrogen-feeding bacterial colony reside inside the catheter site, which break the nitrogen wastes [55].

The most effective way to prevent the CAUTIs, give proper guidance to medical staff and insist the nurse to do more practice [56]. In United Kingdom [57] developed the set rules in the name of "epic3 Guidelines "which based one scientific literature and expertise of medical staff. This booklet proposed that application catheter insertion must be done when there is no alternative because catheter insertion exceed the chance of urinary tract infections [58]. Catheter dressing, sterile catheter bag, length of catheter accordance to patient, gloves and aprons are properly used during catheter-insertion. One most important point to change the urinary drainage bag after every 7 days [55].

2.3 Surgical site infections

After urinary tract infections surgical site infections are most common HCAIs. According to a study about 13% of patients who undergo surgery become infected with Surgical Site Infections and SSI account for about 20% of all Health care associated infections and account for 77% deaths of surgical patients. SSI adopt the pandemic nature; overdose of antibiotic and hospital stay cause the recorded cases in Spain (26.1%) and Europe (19.6%). Reported data shown that SSIs are most common in china, but major microbes associated with SSIs are *E. coli* (25.9%), *S. aureus* (14.3%) and *P. aeruginosa* (11.9%) [25, 59]. Adverse outcomes of SSI include failure of wound healing, increase hospital stay, increase health care cost and mortality. Surgical site infection can occur after days and year of exposure. Center of Disease Control and Prevention (CDC's) classified the SSIs into three major group on the basis of site of infection such as superficial incisional (Skin infection), Deep incisional (Muscle infections) and Organ or Space (any part of body except skin and muscle) [60]. According to WHO resource limited countries like have no more data about surgical site infections. According to an observation study conducted in Pakistan about 6.5% patients who undergoes surgery develop surgical site infections and *Staphylococcus aureus* is most common bacteria that cause SSI. However, *Klebsiellapneumonia*, *Pseudomonas aeruginosa*, *Escherichia coli*, *Acinetobacter* and *Proteus mirabilis* can play significant role in SSI. Time duration of surgery is directly related to infection rate. Caesarian section surgery that last for more than 1.5 hours increases the risk of SSI Infection. Patients with gastrointestinal surgery and wound contamination have high incidence of SSI [61]. Duration of surgery, age of patients, co morbidity and obesity are risk factors for developing surgical site associated infections. Control measures, proper antibiotics prophylaxis, patients' hygienic conditions and good surgery setting can reduce risk of SSI. In ICU Skin and soft tissue infections are most common condition with fatality rate of about 1.3–7.2%. Among 2 million nosocomial infection (20–25%) [62] that occurs every year they account one quarter of these infections [63–65].

A recent study shows that Patients with neurosurgery have evidences of meningitides mostly caused by *Staphylococcus aureus.* Per year number of cases of CHD (Coronary heart disease) is greater which need surgery. Surgery results in many postoperative infections which cause morbidity and mortality of children with CHD. Complexity of surgery, age and contaminations are risk factors for nosocomial infections [64, 66, 67].

Despite of pathophysiological feature of SSIs, government and medical staff should be recommended the preoperative and intraoperative measures to control infections.

*Balanced Diet*plays a critical role in healing of wound if proper nutrient would not take, then it could alter the physiological nature of wound [68]. Proper nutrition boosts up immune response for infection, if a patient should be used "immune-nutrition" [69], it increases the anti-inflammatory response to infection and healing of wound would be rapidly recovered in immunosuppressant patients which suffer with major surgery [70].

Refinement _ Nasal Mupirocin It is a monocarboxylic acid antibacterial agent which is used to stunt the growth of methicillin-resistant *Staphylococcus aureus* (MRSA) [71] specifically in cardiac surgery [72] infections [68, 73].

Immunomodulatory Therapies Inflammatory diseases, Transplant patients and preoperative discontinuations are the risk factor to progression of SSIs [74]. Despite of these factor, scientists are crucial to overcome these limitation in these patient by the application anti-inflammatory drug, methotrexate, which is continuously supplied to patient at preoperative period [75–77].

Bathing/Shower For proper disinfection of skin, chlorhexidine soap [78] and povidone-iodine soap [79] best non-pharmacological soap to eradicate the bacteria but timing, types of soap and number of applications are also mandatory with respect of location such as axilla, groin and skin folds [80].

Oral Antibiotics Consortia of oral antibiotic and bowel [81] is efficient method to reduce the risk of SSIs approximately 4%, specifically in colon surgery [82] which face the exposure of *Clostridium difficile* [83]. This type of antibiotic is very workable on gram-negative bacteria and anaerobes which mutate the surgical complication into severe problem in organs except skin and muscles [84, 85].

Antibiotic Prophylaxis β-lactams is a prophylaxis antibiotic which is used to reduce the chances of SSI in therapeutic tissue [86]. But timing, dosage and indication are the optimal factors to insert in therapeutic tissue. Disproportionate use of antibiotic, which increase toxicity, resistance of bacteria and cost of antibiotics.

Hair Removal from surgical site also enhances the chances of SSIs that's why scientists not recommended hair removal during surgery. But the instrument, which is used to remove hair, really matter for the progression of surgical site infections. Electric shaver, razor blades and depilatory creams are widely used for vigilant hair removal on surgical site [87].

2.4 Ventilator associated pneumonia (VAP)

Ventilator associated pneumonia is one of significant health care issue among health care associated infections. 9–27% patients on ventilators have Ventilator associated pneumonia. 86% of nosocomial pneumonia is ventilator associated. Patients at Intensive Care Unit are more prone to VAP. In Asian countries especially in developing countries incidence of VAP is higher than European countries where poor implementation of standard measures make ICU a major transmitter of Pathogens.

Pseudomonas aeruginosa, Staphylococcus aureus, Acinetobacter and enterococcus are most common causative agents of VAP in Asian countries. Chronic obstructive pulmonary diseases increase incidence of Ventilator associated pneumonia. Old Age, co morbidity, gender and severity of illness are significant risk factors for VAP. Among mechanically ventilated patients in Intensive Care Unit VAP is second most common infection [88]. Studies have shown that critically ill patients on ventilator can also develop nosocomial sinusitis [89].

2.5 Gastroentirites

Inflammation of Gastrointestinal tract is termed as infectious diarrhea or Gastroenteritis. In 2015 globally 1.3 million deaths were reported due to gastroenteritis.

In developing countries prevalence of gastroenteritis is most. Most common causative agent is virus (rotavirus, norovirus, astrovirus and adenovirus) however bacteria, parasites and fungi can also cause gastroenteritis. Most studies in literature show that most of nosocomial gastroenteritis infections were caused by rotavirus and mostly effects children under age of five [90, 91]. According to a study conducted in Pakistan in 2015 about 80% of hospitalized children have viral infections and about 95% were positive for rotavirus in addition to others [92].

2.6 Puerperal fever

During childbirth and after childbirth or miscarriage women get infected with puerperal sepsis. Annually about 75,000 women die worldwide due to puerperal sepsis and developing countries have more death annually than developed countries. Puerperal sepsis is a leading cause of maternal mortality in developing countries like Pakistan due to multiple reasons. Most common causative agent of puerperal fever is bacteria. Data from developing countries as Pakistan shows that more than half of women do not get hospital facilities during delivery. Unhygienic conditions during delivery, long duration of labour, miscarriage, frequent vaginal examination, malnutrition, premature membrane rupturing, and anemia are risk factors for puerperal fever [92]. Most common infection that cause postpartum is endometritis and mostly occur in women who gave birth by cesarean section [93].

3. Causative agents

Bacteria, viruses, and fungus parasites are causative agent responsible for nosocomial infections however most common causative agents are bacteria. In bacteria *Enterobacteria, Staphylococcus and Pseudomonas and Legionella* are more common cause of HCAIs.

80–87% of HCAIs are caused by 12–17 microorganisms *P. aeruginosa, A. baumannii, Enterobacter species, Proteus species, Candida species (eg, albicans, glabrata), K. pneumoniae and Klebsiella oxytoca, E. coli, coagulase-negative Staphylococci, Enterococcus species,* Yeast NOS, Bacteroides species and others. In these 16–20% is multidrug resistance and most of these are gram negative organisms. However causative agents and resistance varies throughout world [94–96]. Bacteria are most common pathogen for HCAIs. Actinobacteria constitutes about 80% infections. Contaminated hands and wounds are mostly affected by Methicillin-resistant *S. aureus* (MRSA) and cause pneumonia and cause surgical site infection [5]. Like bacteria viruses also cause HCAIs. Common viruses causing nosocomial infections are herpes simplex virus, rota virus, influenza, HIV and hepatitis. Fungus such as *Aspergillus sp, Candida albicans, and Cryptococcus neoformans* can cause HCAIs [97]. In addition to bacteria and viruses, fungus (*Aspergillus* and *Candida*), prions and plasmodium can also cause nosocomial infections [98]. Summary of causative agents is shown in **Table 1**.

3.1 Risk factors

Unhealthy hospital environment (poor hygienic conditions, poor medical waste management), unaware medical staff (improper use of invasive devices and medical devices) and susceptibility of patient are risk factors for Health care associated infections (HCAIs).

As these risk factors are mostly associated with poverty so resource limited countries are at more risk to develop HCAIs due to impropriate control policies [98].

DOI: http://dx.doi.org/10.5772/intechopen.97193

Sr. No	Type of HCAIs	Description	Causative agents	Preventions	References
1	Central line -associated bloodstream infections	Fever, tendered site of insertion of IV access of CVP catheter	*Acinetobacter, Candida sp, Citrobacter sp, Corynobacter, E. coli spp, Enterobacter spp, Enterococcus spp, Haemophilusspp, Klebsiella spp, Proteus spp, Pseudomonas spp, Staphylococcus aureus spp, coagulase-negative staphylococci spp, Serratia spp, Stenotrophomonas, Streptococcus*	Hand hygienic, sterilizing techniques, appropriate setting of site, prefer upper extremities for catheter insertion, prefer ultrasound guided insertion, make sure sterile precaution during whole procedure of insertion	[99–102]
2	Catheter associated urinary tract infections	Fever, Lower abdominal pain, changes in urine characteristics	*Staphylococci spp, Alcaligenes denitrificans, Delftiatsuruhatensis, E. coli, Klebsiella pneumoniae, Pseudomonas aeruginosa. D. tsuruhatensi, Serratia marcescens*	Appropriate setting of catheter and site, hand hygienic, sterilizing techniques, sure closed drainage system, sure unobstructed urine flow,	[100–103]
3	Surgical site infections	Fever, wound healing problems, pain, redness	*Staphylococcus aureus, Pseudomonas aeruginosa, Klebsiella pneumonia, E. coli, Acinetobacter,* Coagulase-negative staphylococci	Sterilizing techniques, Safe operating theater, good quality surgical procedure	[104–106]
4	Ventilator associated pneumonia	Decreased intensity of breath sounds, fever, pleuritic chest pain, increase in rales	*Pseudomonas, lebsiella spp., E. coli, Proteus spp, Enterobacter spp, Serratia spp, Citrobacter spp, Streptococcus spp, Hemophilus spp, Acinetobacter spp, Neisseria spp, Stenotrophomonas maltophilia, Coagulase-negative staphylococcus, Corynebacterium, Moraxella, Enterococcus, Nocardia abscesses,* Respiratory syncytial virus, Adenoviruses, Rhinovirus, influenza virus, Herpes simplexvirus	Reduce patient time on ventilator, sterilizing techniques, avoid intubation, elevate head of bed, suction oro pharynx regularly, reduce ventilator circuit changes	[100, 107–110]
5	Gastroenteritis	Increase infrequency of stool, dehydration, fever	Norovirus, Astrovirus Rotavirus, Torovirus, *Adenovirus, Campylobacter, Clostridium difficile,* E.coli, salmonella spp., *Campylobacter jejuni, shigella spp, Staphylococcus aureus, Bacillus sp, Clostridium perfringens* type A, *Clostridium botulinum, Yersinia enterocolitica, Aeromonas spp, Cryptosporidium, Cyclospora cayatenensis, Entamoeba histolytica, Giardia lamblia.*	Use of Probiotics, sterilizing techniques, hand hygiene	[100, 111–118]
6	Puerperal fever	Fever, abdominal distension, wound infection, septicemia and disseminated intravascular coagulation.	*E. coli, Klebsiella pneumoniae, Klebsiella oxytoca, Morganella morganii, Pseudomonas aeruginosa, Coagulase negative staphylococcus, Staphylococcus aureus, Providencia, Corynebacterium sp, Enterobacter sp, Streptococcus pyogenes, Citrobacter freundii, Alcaligenes sp, Shigella sp, Yersinia sp, Streptococcus viridans), streptococcus agalactae, Salmonella sp, Kluyverasp*	Use of sterilized equipment and good health care setting, antiseptic shower after surgery	[119–121]

Table 1.
Summarized causative agents and preventions against different types of HCAIs.

4. Transmission of HAIs

4.1 Hospital environment

Unhealthy hospital setting serves as best source to transmit infections. Contaminated utensils, medical devices, air, food, beds, and windows can transmit pathogens. Supply of filtered air must be maintained in ICU [61, 98].

4.2 Medical staff

Medical staff plays a significant role in prevalence of nosocomial infection. Use of unsterilized medical equipment by unaware medical staff in healthcare delivery increases chances of infection of HCAIs. Improper handling and management of hazardous medical waste by unaware medical staff can act as significant reservoir of HCAIs. Most of studies in Pakistan show non satisfactory behavior of medical staff towards standard precautions [5]. Micro flora of patient can also become source of infection if they effect surgical site or wounds [5].

5. Preventions for HCAIs

5.1 Standard precautions

In health care unit medical staff should adopt proper standard hygienic measures (hand hygienic, sterilized equipment, use of gowns, gloves, respiratory

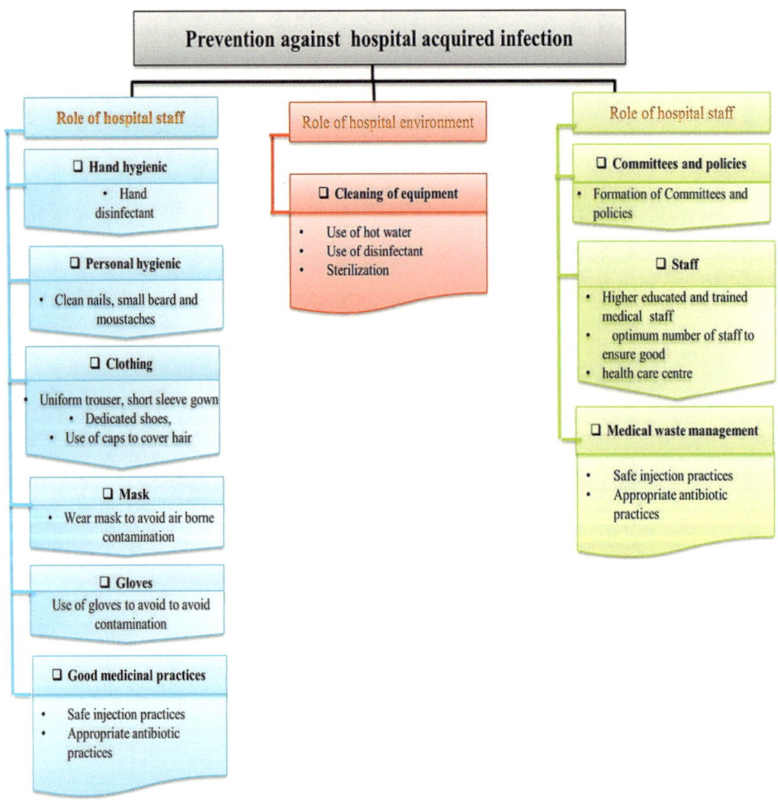

Figure 2.
Prevention against hospital acquired infections.

DOI: http://dx.doi.org/10.5772/intechopen.97193

hygienic) to reduce chances of HCAIs. Medical staff should be trained for biosafety and hazardous waste management should also be maintained. Public should be aware about risk factors consequences of HCAIs as there are number of group of bacteria and viruses in health care centers. Medical staff must be aware with appropriate use of antibiotic to avoid antibiotic resistance which is a significant cause of death in south-East Asian countries where one child died in every five minutes due to antibiotic resistance [5, 97, 122–124]. Preventive measures are the best way to control these type of infections as shown in **Figure 2**.

5.2 Government policies

As HCAIs is leading cause of morbidity and mortality, health institute must plan efficient infection control programs to handle this problem. It is responsibility of government to promote safety of health care centers through availability of trained medical staff, appropriate use of medications and medical equipment and quality eye care. Workload and staff capacity of health care must be directed by government to encourage good health care settings. Government must plan control policies (awareness about HCAIs through media) to reduce risk of Health care associated infections [2].

6. Conclusion

HCAIS is posing serious threat to economy of world specially to developing countries. In resource limited countries infections control program are unsatisfactory. Surveillance for HCAIs mainly serves purpose of prevention interventions. Unhealthy hospital environment and unaware medical staff and susceptibility of patient mainly lead to HCAIs. Government must play its role by forming new policies and committees for modification in national guidelines and for hiring trained and educated staff to promote healthy health care setting. Government should promote implementation of standard strategies by providing resources and policies.

Acknowledgements

Not applicable.

Funding

No funding.

Competing interests

The authors declare that they have no competing interests.

Ethics approval and consent to participate

Not applicable.

Consent for publication

Not applicable.

Availability of data and material

Not applicable.

Authors' contributions

Ayesha Noor and Ali Raza Ishaq both are working as a First author.
All Authors contributed equally.

Author details

Ayesha Noor[1], Ali Raza Ishaq[2], Laila Jafri[3], Faiza Jabeen[4], Rehana Rani[3], Bushra Hafeez Kiani[5], Nosheen Akhtar[6], Zeeshan Javed[1], Tahira Younis[1*] and Fatima Jalal[1,7]

1 Department of Zoology, Government College University Faisalabad, Pakistan

2 State Key Laboratory of Biocatalysis and Enzyme Engineering, Environmental Microbial Technology Center of Hubei Province, College of Life Science, Hubei University, Wuhan, China

3 Department of Life Sciences, Abasyn University Islamabad, Pakistan

4 Department of Zoology, University of Education Lahore, Pakistan

5 Department of Biological Sciences (FC), Faculty of Basic and Applied Sciences, International Islamic University, Islamabad, Pakistan

6 Department of Biological Sciences, National University of Medical Sciences, Rawalpindi, Pakistan

7 Department of Zoology, Government Women University Faisalabad, Pakistan

*Address all correspondence to: tahirayounas@gmail.com

References

[1] Mudassar S, Adeel B, Ali M, Mehmood F, Hussain A. Nosocomial Infections: Awareness and Practices of Nurses Regarding its Spread in a Tertiary Care Hospital of Lahore, Pakistan.

[2] Organization WH. Prevention of hospital-acquired infections: a practical guide. Geneva, Switzerland: World Health Organization; 2002.

[3] Revelas A. Healthcare–associated infections: A public health problem. Nigerian medical journal: journal of the Nigeria Medical Association. 2012;53(2):59.

[4] Shalini S, Vidyasree M, Abiselvi A, Gopalakrishnan S. Impact and effect of nosocomial infections: a review. RESEARCH JOURNAL OF PHARMACEUTICAL BIOLOGICAL AND CHEMICAL SCIENCES. 2015;6(1):947-951.

[5] Khan HA, Baig FK, Mehboob R. Nosocomial infections: Epidemiology, prevention, control and surveillance. Asian Pacific Journal of Tropical Biomedicine. 2017;7(5):478-482.

[6] Hasan R, Jabeen K, Ali A, Rafiq Y, Laiq R, Malik B, et al. Extensively drug-resistant tuberculosis, Pakistan. Emerging infectious diseases. 2010;16(9):1473.

[7] Haque M, Sartelli M, McKimm J, Bakar MA. Health care associated infections–an overview. Infection and drug resistance. 2018;11:2321.

[8] Noakes TD, Borresen J, Hew-Butler T, Lambert M, Jordaan E. Semmelweis and the aetiology of puerperal sepsis 160 years on: an historical review. Epidemiology & Infection. 2008;136(1):1-9.

[9] O'grady NP, Alexander M, Burns LA, Dellinger EP, Garland J, Heard SO, et al. Guidelines for the prevention of intravascular catheter-related infections. Clinical infectious diseases. 2011;52(9):e162-ee93.

[10] Salzman MB, Isenberg HD, Shapiro JF, Lipsitz PJ, Rubin LG. A Prospective Study of the Catheter Hub as the Portal of Entry for Microorganisms Causing Catheter-Related Sepsis in Neonates. The Journal of Infectious Diseases. 1993;167(2):487-490.

[11] Raad I, Costerton W, Sabharwal U, Sadlowski M, Anaissie E, Bodey GP. Ultrastructural analysis of indwelling vascular catheters: a quantitative relationship between luminal colonization and duration of placement. Journal of Infectious Diseases. 1993;168(2):400-407.

[12] Ziegler MJ, Pellegrini DC, Safdar N. Attributable mortality of central line associated bloodstream infection: systematic review and meta-analysis. Infection. 2015;43(1):29-36.

[13] Blot K, Bergs J, Vogelaers D, Blot S, Vandijck D. Prevention of Central Line–Associated Bloodstream Infections Through Quality Improvement Interventions: A Systematic Review and Meta-analysis. Clinical Infectious Diseases. 2014;59(1):96-105.

[14] Stoll BJ. Infections of the neonatal infant. Textbook of pediatrics. 2007:794-811.

[15] Goudie A, Dynan L, Brady PW, Rettiganti M. Attributable cost and length of stay for central line–associated bloodstream infections. Pediatrics. 2014;133(6):e1525-e1e32.

[16] Tokars JI, Richards C, Andrus M, Klevens M, Curtis A, Horan T, et al. The changing face of surveillance for health care—associated infections. Clinical infectious diseases. 2004;39(9):1347-1352.

[17] Mobley RE, Bizzarro MJ, editors. Central line-associated bloodstream infections in the NICU: successes and controversies in the quest for zero. Seminars in perinatology; 2017: Elsevier.

[18] Crnich CJ, Maki DG. The promise of novel technology for the prevention of intravascular device-related bloodstream infection. I. Pathogenesis and short-term devices. Clinical infectious diseases : an official publication of the Infectious Diseases Society of America. 2002;34(9): 1232-1242.

[19] Ishaq AR, Manzoor M, Hussain A, Altaf J, Rehman Su, Javed Z, et al. Prospect of microbial food borne diseases in Pakistan: a review. Brazilian Journal of Biology. 2021;81:940-953.

[20] Aslam S. Effect of antibacterials on biofilms. American journal of infection control. 2008;36(10):S175. e9-S. e11.

[21] Umscheid CA, Mitchell MD, Doshi JA, Agarwal R, Williams K, Brennan PJ. Estimating the proportion of healthcare-associated infections that are reasonably preventable and the related mortality and costs. Infection control and hospital epidemiology. 2011;32(2):101-114.

[22] Couto RC, Carvalho EA, Pedrosa TM, Pedroso ÊR, Neto MC, Biscione FM. A 10-year prospective surveillance of nosocomial infections in neonatal intensive care units. American journal of infection control. 2007;35(3):183-189.

[23] Chun P, Kong S-G, Byun S-Y, Park S-E, Lee H-D. Analysis of neonatal sepsis in one neonatal intensive care unit for 6 years. Korean Journal of Pediatrics. 2010;53(4):495-502.

[24] Haque A, Ahmed S, Rafique Z, Abbas Q, Jurair H, Ali S. Device-associated infections in a paediatric intensive care unit in Pakistan. Journal of Hospital Infection. 2017;95(1):98-100.

[25] Badia J, Casey A, Petrosillo N, Hudson P, Mitchell S, Crosby C. Impact of surgical site infection on healthcare costs and patient outcomes: a systematic review in six European countries. Journal of Hospital Infection. 2017;96(1):1-15.

[26] Menyhay SZ, Maki DG. Disinfection of needleless catheter connectors and access ports with alcohol may not prevent microbial entry: the promise of a novel antiseptic-barrier cap. infection control and hospital epidemiology. 2006;27(1):23-27.

[27] Group UNSS. Patient volume, staffing, and workload in relation to risk-adjusted outcomes in a random stratified sample of UK neonatal intensive care units: a prospective evaluation. The Lancet. 2002;359(9301):99-107.

[28] Krein SL, Kuhn L, Ratz D, Chopra V. Use of designated nurse PICC teams and CLABSI prevention practices among US hospitals: a survey-based study. Journal of patient safety. 2019;15(4):293-295.

[29] Panagiotounakou P, Antonogeorgos G, Gounari E, Papadakis S, Labadaridis J, Gounaris A. Peripherally inserted central venous catheters: frequency of complications in premature newborn depends on the insertion site. Journal of Perinatology. 2014;34(6):461-463.

[30] Breschan C, Platzer M, Jost R, Schaumberger F, Stettner H, Likar R. Comparison of catheter-related infection and tip colonization between internal jugular and subclavian central venous catheters in surgical neonates. Anesthesiology: The Journal of the American Society of Anesthesiologists. 2007;107(6):946-953.

[31] Schulman J, Stricof R, Stevens TP, Horgan M, Gase K, Holzman IR, et al. Statewide NICU central-line-associated bloodstream infection rates decline after

bundles and checklists. Pediatrics. 2011;127(3):436-444.

[32] Kieran EA, O'Sullivan A, Miletin J, Twomey AR, Knowles SJ, O'Donnell CPF. 2% chlorhexidine–70% isopropyl alcohol versus 10% povidone–iodine for insertion site cleaning before central line insertion in preterm infants: a randomised trial. Archives of Disease in Childhood-Fetal and Neonatal Edition. 2018;103(2):F101-F1F6.

[33] Lai NM, Taylor JE, Tan K, Choo YM, Kamar AA, Muhamad NA. Antimicrobial dressings for the prevention of catheter-related infections in newborn infants with central venous catheters. Cochrane Database of Systematic Reviews. 2016(3).

[34] Senese R, Cioffi F, Petito G, Goglia F, Lanni A. Thyroid hormone metabolites and analogues. Endocrine. 2019;66(1):105-114.

[35] Williams FL, Watson J, Day C, Soe A, Somisetty SK, Jackson L, et al. Thyroid dysfunction in preterm neonates exposed to iodine. Journal of Perinatal Medicine. 2017;45(1):135-143.

[36] Foster JP, Richards R, Showell MG, Jones LJ. Intravenous in-line filters for preventing morbidity and mortality in neonates. Cochrane Database of Systematic Reviews. 2015(8).

[37] Jardine LA, Inglis GD, Davies MW. Prophylactic systemic antibiotics to reduce morbidity and mortality in neonates with central venous catheters. Cochrane database of systematic reviews. 2008(1).

[38] Taylor JE, Tan K, Lai NM, McDonald SJ. Antibiotic lock for the prevention of catheter-related infection in neonates. Cochrane Database of Systematic Reviews. 2015(6).

[39] Chenoweth C, Saint S. Preventing catheter-associated urinary tract infections in the intensive care unit. Critical care clinics. 2013;29(1):19-32.

[40] Chenoweth CE, Gould CV, Saint S. Diagnosis, management, and prevention of catheter-associated urinary tract infections. Infectious Disease Clinics. 2014;28(1):105-119.

[41] Nassikas NJ, Monteiro JFG, Pashnik B, Lynch J, Carino G, Levinson AT. Intensive care unit rounding checklists to reduce catheter-associated urinary tract infections. Infection control and hospital epidemiology. 2020;41(6):680-683.

[42] Ghauri SK, Javaeed A, Abbasi T, Khan AS, Mustafa KJ. Knowledge and attitude of health workers regarding catheter-associated urinary tract infection in tertiary care hospitals, Pakistan. JPMA The Journal of the Pakistan Medical Association. 2019;69(12):1843.

[43] Meddings J, Greene MT, Ratz D, Ameling J, Fowler KE, Rolle AJ, et al. Multistate programme to reduce catheter-associated infections in intensive care units with elevated infection rates. BMJ quality & safety. 2020;29(5):418-429.

[44] Bardsley A. Preventing urinary tract infections in catheter care. Nursing And Residential Care. 2017;19(5):260-263.

[45] Kranz J, Schmidt S, Wagenlehner F, Schneidewind L. Catheter-Associated Urinary Tract Infections in Adult Patients. Deutsches Arzteblatt International. 2020;117(6):83-88.

[46] Flores-Mireles AL, Walker JN, Caparon M, Hultgren SJ. Urinary tract infections: epidemiology, mechanisms of infection and treatment options. Nature reviews microbiology. 2015;13(5):269-284.

[47] Nimri L, Sulaiman M, Hani OB. Community-acquired urinary tract infections caused by *Burkholderia cepacia* complex in patients with no underlying risk factor. JMM case reports. 2017;4(1).

[48] Bagchi S, Watkins J, Norrick B, Scalise E, Pollock DA, Allen-Bridson K. Accuracy of catheter-associated urinary tract infections reported to the National Healthcare Safety Network, January 2010 through July 2018. American journal of infection control. 2020;48(2):207-211.

[49] O'brien VP, Hannan TJ, Nielsen HV, Hultgren SJ. Drug and vaccine development for the treatment and prevention of urinary tract infections. Urinary Tract Infections: Molecular Pathogenesis and Clinical Management. 2017:589-646.

[50] Tambyah PA, Maki DG. Catheter-associated urinary tract infection is rarely symptomatic: a prospective study of 1497 catheterized patients. Archives of internal medicine. 2000;160(5): 678-682.

[51] Levison ME, Kaye D. Treatment of complicated urinary tract infections with an emphasis on drug-resistant gram-negative uropathogens. Current infectious disease reports. 2013;15(2): 109-115.

[52] Pallett A, Hand K. Complicated urinary tract infections: practical solutions for the treatment of multiresistant Gram-negative bacteria. Journal of antimicrobial chemotherapy. 2010;65(suppl_3):iii25-iii33.

[53] Aaronson DS, Wu AK, Blaschko SD, McAninch JW, Garcia M. National incidence and impact of noninfectious urethral catheter related complications on the Surgical Care Improvement Project. The Journal of urology. 2011;185(5):1756-1760.

[54] Nicolle LE. Catheter-related urinary tract infection. Drugs & aging. 2005;22(8):627-639.

[55] Loveday HP, Wilson JA, Pratt RJ, Golsorkhi M, Tingle A, Bak A, et al. epic3: national evidence-based guidelines for preventing healthcare-associated infections in NHS hospitals in England. Journal of Hospital Infection. 2014;86:S1-S70.

[56] Rinke ML, Oyeku SO, Heo M, Saiman L, Zachariah P, Rosenberg RE, et al. Pediatric ambulatory catheter-associated urinary tract infections (CAUTIs): Incidence, risk factors, and patient outcomes. Infection control and hospital epidemiology. 2020:1-9.

[57] Shah SM, Hussain F. Ethnomedicinal plant wealth of Mastuj valley, Hindukush range, District Chitral, Pakistan. Journal of Medicinal Plants Research. 2012;6(26):4328-4337.

[58] Gould D. Preventing catheter-associated urinary tract infection. Nursing Standard (2014+). 2015;30 (10):50.

[59] Fan Y, Wei Z, Wang W, Tan L, Jiang H, Tian L, et al. The incidence and distribution of surgical site infection in mainland China: a meta-analysis of 84 prospective observational studies. Scientific reports. 2014;4:6783.

[60] Control CfD, Prevention. National Healthcare Safety Network: surgical site infection (SSI) event. 2010.

[61] Owens C, Stoessel K. Surgical site infections: epidemiology, microbiology and prevention. Journal of hospital infection. 2008;70:3-10.

[62] Merchant R. Infectious disease. Update on emerging infections: news from the Centers for Disease Control and Prevention. Annals of emergency medicine. 2011;58(1):67-68.

[63] Murad HF, Inam Pal KM. Nosocomial infections in the ICU: Pens and spectacles as fomites. JPMA: Journal of Pakistan Medical Association. 2016;66(10):S-53.

[64] Sattar F, Sattar Z, Mohsin Zaman SA. Frequency of post-operative surgical site infections in a Tertiary care hospital in Abbottabad, Pakistan. Cureus. 2019;11(3).

[65] Mangram AJ, Horan TC, Pearson ML, Silver LC, Jarvis WR. Guideline for Prevention of Surgical Site Infection, 1999. Centers for Disease Control and Prevention (CDC) Hospital Infection Control Practices Advisory Committee. American journal of infection control. 1999;27(2):97-132; quiz 3-4; discussion 96.

[66] Weisfelt M, Van de Beek D, Spanjaard L, De Gans J. Nosocomial bacterial meningitis in adults: a prospective series of 50 cases. Journal of Hospital Infection. 2007;66(1):71-78.

[67] Sharan H, Misra AP, Mishra R. Determinants of surgical site infection in rural Kanpur, India. J Evol Med Dent Sci. 2012;1(6):921-928.

[68] Allegranzi B, Bischoff P, de Jonge S, Kubilay NZ, Zayed B, Gomes SM, et al. New WHO recommendations on preoperative measures for surgical site infection prevention: an evidence-based global perspective. The Lancet Infectious Diseases. 2016;16(12):e276-ee87.

[69] Howes N, Atkinson C, Thomas S, Lewis SJ. Immunonutrition for patients undergoing surgery for head and neck cancer. Cochrane Database of Systematic Reviews. 2018(8).

[70] Xu J, Sun X, Xin Q, Cheng Y, Zhan Z, Zhang J, et al. Effect of immunonutrition on colorectal cancer patients undergoing surgery: a meta analysis. International journal of colorectal disease. 2018;33(3):273-283.

[71] Tucaliuc A, Blaga AC, Galaction AI, Cascaval D. Mupirocin: applications and production. Biotechnology letters. 2019;41(4-5):495-502.

[72] Kalmeijer M, Coertjens H, van Nieuwland-Bollen P, Bogaers-Hofman D, de Baere GJ, Stuurman A, et al. Surgical site infections in orthopedic surgery: the effect of mupirocin nasal ointment in a double-blind, randomized, placebo-controlled study. Clinical Infectious Diseases. 2002;35(4):353-358.

[73] García AM, Villa MV, Escudero ME, Gómez P, Vélez MM, Múnera MI, et al. Use of nasal mupirocin for Staphylococcus aureus: effect on nasal carriers and nosocomial infections. Biomedica. 2003;23(2):173-179.

[74] Tablan O. Healthcare Infection Control Practices Advisory Committee, Centers for Disease Control and Prevention. Guidelines for preventing health-care-associated pneumonia, 2003: recommendations of the CDC and the Healthcare Infection Control Practices Advisory Committee. MMWR Recomm Rep. 2004;53(3):1-36.

[75] Berthold E, Geborek P, Gülfe A. Continuation of TNF blockade in patients with inflammatory rheumatic disease. An observational study on surgical site infections in 1,596 elective orthopedic and hand surgery procedures. Acta orthopaedica. 2013;84(5):495-501.

[76] Saitoh M, Matsushita K. [Prevention of surgical site infection for orthopaedic surgery in rheumatoid arthritis]. Nihon rinsho Japanese journal of clinical medicine. 2016;74(6):993-999.

[77] Wang AS, Armstrong EJ, Armstrong AW. Corticosteroids and wound healing: clinical considerations in the perioperative period. The American journal of surgery. 2013;206(3):410-417.

[78] Garibaldi RA, Skolnick D, Lerer T, Poirot A, Graham J, Krisuinas E, et al. The Impact of Preoperative Skin

Disinfection on Preventing Intraoperative Wound Contamination. Infection Control & Hospital Epidemiology. 1988;9(3):109-113.

[79] Leigh D, Stronge J, Marriner J, Sedgwick J. Total body bathing with 'Hibiscrub'(chlorhexidine) in surgical patients: a controlled trial. Journal of Hospital Infection. 1983;4(3):229-235.

[80] Kamel C, McGahan L, Polisena J, Mierzwinski-Urban M, Embil JM. Preoperative skin antiseptic preparations for preventing surgical site infections: a systematic review. infection control and hospital epidemiology. 2012;33(6):608.

[81] Zmora O, Mahajna A, Bar-Zakai B, Rosin D, Hershko D, Shabtai M, et al. Colon and rectal surgery without mechanical bowel preparation: a randomized prospective trial. Annals of surgery. 2003;237(3):363.

[82] Lewis RT. Oral versus systemic antibiotic prophylaxis in elective colon surgery: a randomized study and meta-analysis send a message from the 1990s. Canadian journal of surgery. 2002;45(3):173.

[83] Nelson RL, Gladman E, Barbateskovic M. Antimicrobial prophylaxis for colorectal surgery. Cochrane Database of Systematic Reviews. 2014(5).

[84] Englesbe MJ, Brooks L, Kubus J, Luchtefeld M, Lynch J, Senagore A, et al. A statewide assessment of surgical site infection following colectomy: the role of oral antibiotics. Annals of surgery. 2010;252(3):514.

[85] Althumairi AA, Canner JK, Pawlik TM, Schneider E, Nagarajan N, Safar B, et al. Benefits of bowel preparation beyond surgical site infection. Annals of surgery. 2016;264(6):1051-1057.

[86] Anderson DJ, Podgorny K, Berrios-Torres SI, Bratzler DW, Dellinger EP, Greene L, et al. Strategies to prevent surgical site infections in acute care hospitals: 2014 update. Infection Control & Hospital Epidemiology. 2014;35(S2):S66-S88.

[87] Kluytmans JA. Nasal carriage of Staphylococcus aureus is a major risk factor for surgical-site infections in orthopedic surgery. Infection Control and Hospital Epidemiology. 2000.

[88] Usman HS, Atif I, Rashid F, Zulfiqar H, Mian K, Sarfraz M, et al. Knowledge and practices of critical care health professionals related to ventilator associated pneumonia in tertiary care hospitals of Islamabad and Rawalpindi. JPMA The Journal of the Pakistan Medical Association. 2017;67(11): 1714-1718.

[89] Aggarwal S, Azim A, Baronia A, Kumar R. Evaluation and management of nosocomial sinusitis in Intensive Care Unit patients for pyrexia of unknown origin: Case report and review of literature. International Journal of Medicine and Biomedical Research. 2012;1(2):161-166.

[90] Tate JE, Burton AH, Boschi-Pinto C, Steele AD, Duque J, Parashar UD. 2008 estimate of worldwide rotavirus-associated mortality in children younger than 5 years before the introduction of universal rotavirus vaccination programmes: a systematic review and meta-analysis. The Lancet infectious diseases. 2012;12(2):136-141.

[91] Webber R. Communicable disease epidemiology and control: a global perspective: Cabi; 2009.

[92] Alam MM, Khurshid A, Shaukat S, Rana MS, Sharif S, Angez M, et al. Viral etiologies of acute dehydrating gastroenteritis in pakistani children: confounding role of parechoviruses. Viruses. 2015;7(1):378-393.

[93] Madhudas C, Khurshid F, Sirichand P. Maternal morbidity and mortality associated with puerperal sepsis. Journal of Liaquat University of Medical and Health Sciences. 2011;10(03):121.

[94] Sievert DM, Ricks P, Edwards JR, Schneider A, Patel J, Srinivasan A, et al. Antimicrobial-resistant pathogens associated with healthcare-associated infections: summary of data reported to the National Healthcare Safety Network at the Centers for Disease Control and Prevention, 2009-2010. Infection control and hospital epidemiology. 2013;34(1):1-14.

[95] Weiner LM, Webb AK, Limbago B, Dudeck MA, Patel J, Kallen AJ, et al. Antimicrobial-resistant pathogens associated with healthcare-associated infections: summary of data reported to the National Healthcare Safety Network at the Centers for Disease Control and Prevention, 2011-2014. infection control & hospital epidemiology. 2016;37(11): 1288-301.

[96] Hidron AI, Edwards JR, Patel J, Horan TC, Sievert DM, Pollock DA, et al. Antimicrobial-resistant pathogens associated with healthcare-associated infections: annual summary of data reported to the National Healthcare Safety Network at the Centers for Disease Control and Prevention, 2006-2007. Infection control and hospital epidemiology. 2008;29(11): 996-1011.

[97] Ducel G, Fabry J, Nicolle L. Prevention of hospital acquired infections: a practical guide. Prevention of hospital acquired infections: a practical guide. 2002(Ed. 2).

[98] Chaudhry S, Hussain R. Postpartum infection can be a disaster. Pak J Med Dent. 2014;3(4):70-73.

[99] Leblebicioglu H, Öztürk R, Rosenthal VD, Akan ÖA, Sirmatel F, Ozdemir D, et al. Impact of a multidimensional infection control approach on central line-associated bloodstream infections rates in adult intensive care units of 8 cities of Turkey: findings of the International Nosocomial Infection Control Consortium (INICC). Annals of clinical microbiology and antimicrobials. 2013;12(1):1-10.

[100] Mbim EN, Mboto CI, Agbo BE. A review of nosocomial infections in sub-Saharan Africa. Microbiology Research Journal International. 2016:1-11.

[101] Dumont C, Nesselrodt D. Preventing central line-associated bloodstream infections CLABSI. Nursing2019. 2012;42(6):41-6.

[102] Mehta Y, Gupta A, Todi S, Myatra S, Samaddar D, Patil V, et al. Guidelines for prevention of hospital acquired infections. Indian journal of critical care medicine: peer-reviewed, official publication of Indian Society of Critical Care Medicine. 2014;18(3):149.

[103] Majumder MMI, Ahmed T, Ahmed S, Khan AR. Microbiology of Catheter Associated Urinary Tract Infection. Microbiology of Urinary Tract Infections-Microbial Agents and Predisposing Factors: IntechOpen; 2018.

[104] Mundhada AS, Tenpe S. A study of organisms causing surgical site infections and their antimicrobial susceptibility in a tertiary care government hospital. Indian Journal of Pathology and Microbiology. 2015;58(2):195.

[105] Nasser A, Zhang X, Yang L, Sawafta FJ, Salah B. Assessment of surgical site infections from signs & symptoms of the wound and associated factors in public hospitals of Hodeidah City, Yemen. Int J Appl. 2013;3(3): 101-110.

[106] Spagnolo A, Ottria G, Amicizia D, Perdelli F, Cristina ML. Operating theatre quality and prevention of surgical site infections. Journal of preventive medicine and hygiene. 2013;54(3):131.

[107] Kalanuria AA, Zai W, Mirski M. Ventilator-associated pneumonia in the ICU. Critical care. 2014;18(2):208.

[108] Park DR. The microbiology of ventilator-associated pneumonia. Respiratory care. 2005;50(6):742-765.

[109] Daubin C, Vincent S, Vabret A, du Cheyron D, Parienti J-J, Ramakers M, et al. Nosocomial viral ventilator-associated pneumonia in the intensive care unit: a prospective cohort study. Intensive care medicine. 2005;31(8):1116-1122.

[110] Huang Y-T, Huang C-Y, Su H-Y, Ma C-T. Using TRM to Enhance the Accuracy of Ventilator-Associated Pneumonia Preventive Measures Implemented by Neonatal Intensive Care Unit Medical Staffs. Hu Li Za Zhi. 2018;65(3):71-79.

[111] Lopman BA, Reacher MH, Vipond IB, Hill D, Perry C, Halladay T, et al. Epidemiology and cost of nosocomial gastroenteritis, Avon, England, 2002-2003. Emerging infectious diseases. 2004;10(10):1827.

[112] Jamieson FB, Wang EE, Bain C, Good J, Duckmanton L, Petric M. Human torovirus: a new nosocomial gastrointestinal pathogen. The Journal of infectious diseases. 1998;178(5): 1263-1269.

[113] Pang XL, Vesikari T. Human astrovirus-associated gastroenteritis in children under 2 years of age followed prospectively during a rotavirus vaccine trial. Acta Paediatrica. 1999;88(5): 532-536.

[114] Chandra BK, Singh G, Taneja N, Pahil S, Singhi S, Sharma M. Diarrhoeagenic Escherichia coli as a predominant cause of paediatric nosocomial diarrhoea in India. Journal of medical microbiology. 2012;61(6):830-836.

[115] Lam B, Tam J, Ng M, Yeung C. Nosocomial gastroenteritis in paediatric patients. Journal of Hospital Infection. 1989;14(4):351-355.

[116] Bobo LD, Dubberke ER. Recognition and prevention of hospital-associated enteric infections in the intensive care unit. Critical care medicine. 2010;38(8 0):S324.

[117] Sideroglou T, Kontopidou F, Mellou K, Maragos A, Potamiti-Komi M, Gerakis T, et al. Management and investigation of viral gastroenteritis nosocomial outbreaks: lessons learned from a recent outbreak, Greece, 2012. Hippokratia. 2014;18(3):204.

[118] Szajewska H, Guarino A, Hojsak I, Indrio F, Kolacek S, Shamir R, et al. Use of probiotics for management of acute gastroenteritis: a position paper by the ESPGHAN Working Group for Probiotics and Prebiotics. Journal of Pediatric Gastroenterology and Nutrition. 2014;58(4):531-539.

[119] Majangara R, Gidiri MF, Chirenje ZM. Microbiology and clinical outcomes of puerperal sepsis: a prospective cohort study. Journal of Obstetrics and Gynaecology. 2018;38(5):635-641.

[120] Khaskheli M-N, Baloch S, Sheeba A. Risk factors and complications of puerperal sepsis at a tertiary healthcare centre. Pakistan journal of medical sciences. 2013;29(4):972.

[121] Huber CP. Prevention and treatment of puerperal sepsis: Baird, D., Michie, A., and MacDonald, R.: Lancet 2: 148, 1939. American Journal of Obstetrics & Gynecology. 1941;41(1):171.

[122] Pittet D, Allegranzi B, Storr J, Nejad SB, Dziekan G, Leotsakos A, et al. Infection control as a major World Health Organization priority for developing countries. Journal of Hospital Infection. 2008;68(4):285-292.

[123] Control CfD, Prevention. Diseases and organisms in healthcare settings. Secondary diseases and organisms in healthcare settings. 2016.

[124] Leekha S, Terrell CL, Edson RS, editors. General principles of antimicrobial therapy. Mayo Clinic Proceedings; 2011: Elsevier.